Purification of
Fermentation Products

ACS SYMPOSIUM SERIES 271

Purification of Fermentation Products

Applications to Large-Scale Processes

Derek LeRoith, EDITOR
National Institutes of Health

Joseph Shiloach, EDITOR
National Institutes of Health

Timothy J. Leahy, EDITOR
Millipore Corporation

Based on a symposium sponsored by
the Division of Microbial and Biochemical Technology

American Chemical Society, Washington, D.C. 1985

Library of Congress Cataloging in Publication Data
Purification of fermentation products.
  (ACS symposium series, ISSN 0097-6156; 271)

  Includes bibliographies and indexes.

  1. Fermentation—Congresses. 2. Separation
(Technology)—Congresses.

  I. LeRoith, Derek, 1945–   . II. Shiloach, Joseph.
III. Leahy, Timothy J., 1949–   . IV. American
Chemical Society. Division of Microbial and
Biochemical Technology. V. Series.

TP156.F4P87   1985   660.2′8449   84–24316
ISBN 0–8412–0890–5

T P 1 5 6
F 4 P 8 7
1 9 8 5
C H E M

ACS Symposium Series

M. Joan Comstock, *Series Editor*

Advisory Board

FOREWORD

The ACS Symposium Series was founded in 1974 to provide a medium for publishing symposia quickly in book form. The format of the Series parallels that of the continuing Advances in Chemistry Series except that in order to save time the papers are not typeset but are reproduced as they are submitted by the authors in camera-ready form. Papers are reviewed under the supervision of the Editors with the assistance of the Series Advisory Board and are selected to maintain the integrity of the symposia; however, verbatim reproductions of previously published papers are not accepted. Both reviews and reports of research are acceptable since symposia may embrace both types of presentation.

CONTENTS

PREFACE

THE ADVENT OF GENETIC ENGINEERING through the use of recombinant DNA techniques has stimulated wide interest in creating useful biologically derived products. There exists today a significant family of products that has been traditionally synthesized through the action of microorganisms. Typically, the manufacture of a biologically derived product begins with mass cultivation of the desired microorganism, a process known as fermentation.

Fermentation, however, represents only one step in the overall scheme of product formation. It is equally important to recover and purify the desired product from the fermentation after the organism has been cultivated. Often, that product is present in relatively low concentrations compared to the other components of the fermentation, and the challenge lies in separating what is wanted from what is not in an efficient, economical, and timely fashion.

Although great strides have been made in the genetic manipulation of microorganisms, a concomitant increase in new product-recovery methods has lagged behind. Recently, though, many groups of researchers in both academia and industry have initiated active development programs in the area of fermentation product recovery and purification.

This book presents some of these newer approaches to product recovery. The emphasis is on large-scale processes where several approaches to product isolation are required to obtain pure material. Each author discusses state-of-the-art approaches to product purification that uniquely fit the material to be isolated.

Two separation technologies are highlighted here: filtration and chromatography. Both technologies have existed for some time, but only recently have they found wider use in fermentation processes. The chapters in this book describe some of the newest applications of these technologies to product purification. The authors provide a cross section of viewpoints from academia and industry and consider both the practical and theoretical aspects of biological processing.

We wish to thank all of the authors for their contributions to this book. We also thank the Division of Microbial and Biochemical Technology, and specifically, Dr. Larry Robertson, for sponsorship of this work. Finally, we wish to express our appreciation to the editorial staff of the ACS Books Department with special thanks to Robin Giroux.

DEREK LEROITH
National Institutes of Health

JOSEPH SHILOACH
National Institutes of Health

TIMOTHY J. LEAHY
Millipore Corporation

October 5, 1984

Processing Cell Lysate with Tangential Flow Filtration

RAYMOND GABLER and MARY RYAN

Millipore Corporation, Bedford, MA 01730

Operating conditions and the corresponding performance
of a membrane filter system have been examined for
recovering proteins from a bacterial lysate. Specifically, the dependence of membrane flux on average trans
membrane pressure and recirculation rate has been investigated for both whole cells and lysate. The recovery of an intracellular protein was simulated by adding
a marker protein, IgG, to the lysate after cell lysis.
Protein concentrations were measured at each step of
the processing scheme to determine the distribution of
IgG and how much could be recovered. Operating parameters that influence the flow of IgG through a microporous membrane have also been studied.

Since the introduction of genetic engineering on a practical scale
in the 1970's, there has been increased interest in the production
of biological products using large scale fermentation. Both new
products that had not been commercially available previously in
sufficient quantities, and old products whose production costs are
now potentially cheaper have been in the spotlight. One of the
difficulties that is encountered with some types of recombinantly
derived proteins, however, is the purification steps. Because E.
coli does not secrete recombinant proteins into the growth media,
the cells must be lysed which increases the difficulty of separating
the desired protein from all other biological constituents. Purification of a protein from a lysate requires a number of steps to
first eliminate cell fragments and debris from soluble material and
to then separate the desired protein from other soluble components.
Techniques that have been classically used in lysate processing and
protein isolation include: centrifugation, open column chromatography and precipitation. Tangential flow filtration can now be
added to the list.[1]
　　Tangential flow filtration has been most extensively used for
concentrating cells[2,3], concentrating and washing proteins[4] and
removing pyrogens from solutions[5,6]. By passing the process
solution tangential or parallel to the filter, and recirculating this
fluid back to the original container, it is possible to filter solu-

tions that would normally clog a dead ended filtration system. Also, with filtration, it is possible to perform gross fractionations based on size. These two factors make it feasible to process cell lysates with membrane filters in order to separate cell debris from proteins, and to also concentrate the crude fraction prior to the next purification step. Filtration offers advantages over other techniques in that relatively little energy is needed compared to centrifugation, and the filtration system is closed and will not produce aerosols. Also, the equipment can be scaled up to handle large quantities.

The purpose of this work is several fold. First, we wanted to determine the magnitude of filtration rates that are possible with lysates compared to whole cells, and to identify those operational parameters that influence membrane flux. Second, an effort was made to investigate how several different methods of generating a lysate suspension would affect membrane processing, if at all. Third, we wanted to investigate the quantitatively recovery of a specific protein from a lysate with filtration, and last, an effort was made to determine what operational parameters are most important in maximizing the flow of protein through the filter? In short, this work was designed as the first step in documenting the performance of membrane systems for processing lysates.

To answer the above issues, a model system was devised in which a specific protein was added to an E. coli lysate. The lysate was then processed through the filtration steps and the protein recovered. The protein chosen was human IgG which has a relatively large molecular weight (160,000 Daltons). It was reasoned that if a large protein could be separated from cell debris and passed through a membrane and recovered, then a smaller protein that is typical of a recombinant process, should prove to be much easier.

Materials and Methods

Fermentations. E. coli was grown in a defined media with the following composition:

PO_4^{-2}	0.05M
$MgSO_4$	$10^{-3}M$
$CaCl_2$	$10^{-4}M$
$FeSO_4$	$10^{-5}M$
$(NH_4)_2 SO_4$	0.2%
Glucose	0.5%

The antifoam used was 2% octadecanol in mineral oil (0.5ml/liter of fermentation broth). The fermentors were batched in 11 liter quantities, and growth lasted about 24 hours at 37°C. Rotor speed was 400 RPM and the aeration rate was 5 liters/minute. A Microgen (New Brunswick Scientific) or a Chemap 14 liter fermentor was used. Final viable counts were typically greater than $10^9/ml$.

Membrane Separations. The membrane filter separation system

(Millipore Corporation, XX8140000) consisted of a membrane holder (Pellicon), a 4 gallon per minute rotary vane pump (Procon) and the membranes themselves. The filters were either 0.45 micron microporous (Durapore) or 100,000 NMWL ultrafiltration membranes. All tubing and connections were 1/2 inch. A manifold flow bypass was attached to the pump so fluid could be introduced into the filtration system without a sudden surge of pressure buildup (Bulletin AB822, Millipore Corporation). In all cases, 5 square feet of membrane were used.

Figure 1 shows a schematic of the equipment used and how it was plumbed together. All together there were three modes of operation used for various applications reported here. For concentrating cells or lysate, the system was operated exactly as shown in Figure 1. Filtrate was collected separately while the retentate was circulated back to the original holding tank. In the total recycle mode of operation, the filtrate line was placed in the cell or lysate holding reservoir so the cells or lysate concentration remained constant as a function of time. The reason for performing total recycle is to isolate process variables in determining the independent influence of trans membrane pressure or recirculation rate on flow through the membrane. A constant volume wash mode of operation is accomplished by continuously adding wash buffer to the cell or lysate suspension at the same rate that filtrate is collected. For all modes of operation, there was no filtrate back pressure.

The inlet pressure for the filter holder is measured on the upstream side of the membrane as the fluid enters. The outlet pressure is also monitored on the upstream side of the membrane as the retentate exits the filter holder device. The difference between the inlet and outlet pressure is proportional to the circulation rate while the sum of the inlet and outlet pressures is proportional to the average trans membrane pressure.

Cell Lysis. Cell lysis was accomplished either enzymatically with lysozyme or with sonication.

Sonication. Fermentor broths were reduced in volume via filtration to several liters or less. This reduced volume was then continuously circulated through a sonifer (Branson Cell Disrupter 185). Water from an ice bath was passed through the jacket to dissipate heat generated from the sonifer. Lysis was checked visually by monitoring the number of whole cells seen in a wet mount under an optical microscope at 1000X magnification.

Lysozyme. After fermentation, the cell broth was reduced in volume via filtration to one liter or less. The concentrated cells were then pelleted and washed with 0.85% saline. The washed cells were resuspended in 0.1M EDTA, pH 8.0 and allowed to sit at room temperature for about 45 minutes with gentle stirring. Lysozyme was then added to enough 0.5M NaCl, pH 8.0, to bring the total volume to 1/20th of the original culture volume. The lysozyme concentration was 2 mg/ml in this suspension. The mixture was suspended in 0.1% sodium desoxycholate to finalize the lysis. This procedure[7] produced a very viscous suspension. The viscosity was reduced by adding either streptomycin sulfate or DNAse. In either case, the resulting suspension was allowed to sit overnight with gentle stirring.

IgG, DNAse. Human, freeze dried IgG was purchased from Sigma Chemical (cat. no. HG-11) as Cohn fraction II. The IgG was resuspended
in physiological saline or cell concentration filtrate and the IgG
was added to the lysed cells to a final concentration of around 0.1%-
0.2%. DNAse was also purchased from Sigma Chemical.

IgG Assay. Samples and controls to be assayed for IgG were collected
in 5 ml quantities. From these samples, 750 microliters were centrifuged for 10 minutes at 1250 RPM. The supernant fluid was collected,
and 250 microliters were injected into a Beckman ICS Analyzer II.
The analyzer would mix appropriate amounts of anti IgG antibody with
the sample automatically and measure the change in light scattering
intensity as the antigen-antibody reaction progressed. The formation
of the immunoprecipitin complex proceeds gradually at first, then
rapidly and finally progresses through a peak value. If the sample
IgG is within the correct concentration range, the final peak rate
for the change in light scattering intensity is proportional to the
sample IgG concentration. Normal human sera of known IgG concentration was used for calibration and controls. With lysate samples,
this assay technique was reproducible to about 10%.

Results

A schematic illustrating how membrane filters are used to process
lysates is shown in Figure 2. The overall purpose of the processing
is to separate an intracellular product, typically a protein, from
the bulk of large cell fragments or debris, and to concentrate the
protein to a small workable volume. Basically, there are four
filtration steps in this process. After the cells are grown, the
cells are concentrated to a relatively small volume. This is accomplished in the first filtration step. After lysis, the lysate is
concentrated and washed in the second and third steps respectively.
Washing is performed in order to enhance the passage of protein
through the membrane if the protein remains in the retentate during
the concentration of the lysate. Both concentration and washing can
use the exact same microporous membrane. In the last step, an ultrafiltration membrane is used to concentrate the crudely fractionated
protein solution in preparation for the next purification step which
is usually a chromatographic process of one sort or another.
      The information presented in the following figures represents
two types of results. First, there is the evidence for IgG recovery
from membrane filtration as a function of different biochemical processing schemes. Second, there is the data which describes the flow
rates of fluid or the flux through the membrane as a function of
different operating conditions.
      Figure 3 is a schematic representing a filtration process of
cells, lysate and protein solution in which the lysate was formed by
sonication. Initially, 22.3 liters or cells were concentrated down
to 4.8 liters with 0.45 micron microporous membrane. The cells were
lysed and the IgG added in the amount of 11.1 grams (6.8 liters
total lysate). The lysate was then concentrated to 1.3 liters.
During the lysate concentration, 7.6 grams of IgG passed through the
0.45 micron pore size membrane and 2.5 grams remained in the retentate. Respective fluid volumes were 5.5 liters and 1.3 liters. A
constant volume wash was performed on the 1.3 liters of retentate in

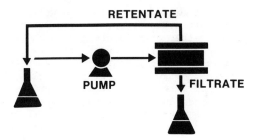

Figure 1. Schematic of equipment for processing lysates. The tangential flow filter is in the stacked sheet configuration which allows high inlet pressure.

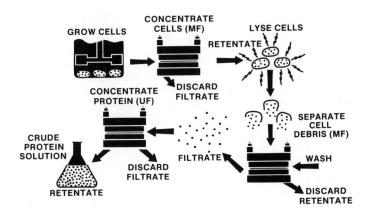

Figure 2. Filtration steps for processing lysate. All steps, except the protein concentration, use a microporous membrane. Protein concentration is accomplished by a UF membrane.

an effort to separate the remaining 2.5 grams of IgG from the cell
debris. After 5 liters of wash was collected in the filtrate, 2.4
grams of IgG had also passed the membrane showing that the IgG can be
recovered and effectively separated from the lysate. The 5.5 liters
from the original lysate concentrate was then concentrated with a
100,000 MWCO UF membrane. In the concentration, only 5.2 grams of
IgG was recovered, however, none of the IgG passed the UF membrane.

In an effort to determine if the loss of IgG in the UF concentration of the crude protein fraction shown in Figure 3 represented
a real issue or just a matter of collecting all the retentate from
the dead volume, the concentration step was repeated. This time,
particular attention was paid to flushing out the retentate hoses and
dead spots in the system. Figure 4 shows the results of this work.
The lysate filtrate (10.5 liters, 7.2 grams of IgG) was split into
two fractions of 5.9 liters and 4.8 liters respectively and concentrated under two separate operating conditions. Operating at an inlet pressure of 80 psi and concentrating the 5.9 liters to 1.4 liters,
3.7 grams of IgG was recovered immediately which represents 95% of
that expected. Subsequently, 1.5 liters of physiological saline was
recirculated through the filtration system for 5 minutes and then
assayed for the remaining IgG. Recovery of another 0.2 grams of IgG
(remaining 5%) was found, so together with the initial 3.7 grams of
IgG, the total amount expected was recovered. For operating conditions of 40 psi inlet pressure and 0 outlet pressure, 88% of the IgG
was recovered. Results from Figure 4 indicate that under both high
and moderate inlet pressures, the IgG could be recovered from the
concentration step.

Figure 5 shows the flow decay for the cell suspension concentration step of Figure 3. The flow rate decays gradually with time with
an average final rate around 700 ml/min. Inlet pressure was 90 psi,
and there was a 4.6X reduction in volume in about 15 minutes. The
flow decay curve in Figure 5 is typical for an E. coli concentration
with a new 0.45 micron pore size microporous membrane.

Figure 6 is the corresponding flow decay curve for the sonified
lysate suspension. Again, this is a typical flow decay with time.
The flux for the lysate is less than that seen for the whole cell
concentration. The flow rate, however, is still significant and the
equilibrium rate corresponds to 23 gallons per square foot per day
(GFD). In other experiments, flow rates for lysate concentration are
also seen to be relatively constant showing little decay.

Figure 7 is a plot of the flow rate over time for the constant
volume wash step used to pass the IgG remaining in the retentate
after the initial concentration of the lysate. As expected, with the
lysate volume remaining constant at about 1.3 liters, there is little
flow decay with time. The arrows in Figure 7 indicate the commence-
ment of the addition of one liter aliquots of saline solution at a
rate matching that of the filtrate. Five square feet of an 0.45
micron pore size filter was used for this procedure.

Figure 8 shows the respective flow decays for the crude protein
fraction concentration with the 100,000 MWCO UF membrane. Inlet
pressures were 80 psi and 40 psi respectively. The higher inlet
pressure gave a higher flux compared to the 40 psi inlet pressure as
would be expected. When the inlet pressure is increased, both the
average TMP and also the recirculation rate increase. Based on the
data in Figure 8 alone, it is not possible to determine if the higher

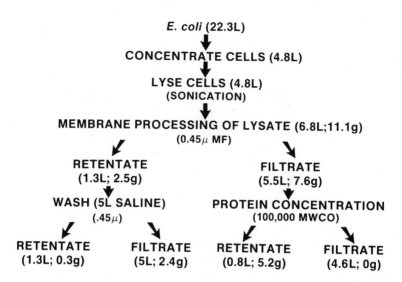

Figure 3. Distribution of IgC for the different process steps. Both the fluid volumes and the total IgC content are given for all retentates and filtrates throughout the whole sequence.

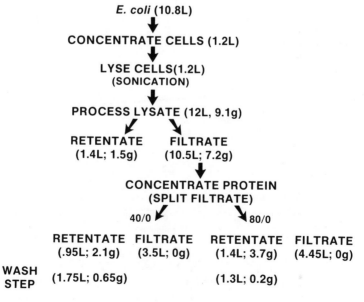

Figure 4. Concentration of protein under different operating conditions. IgC can be recovered quantitatively under two sets of operating pressures by a UF membrane.

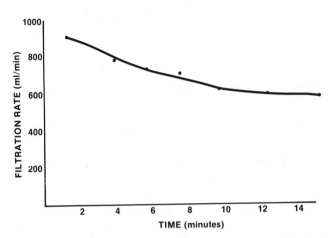

Figure 5.   Flow decay curve for concentrating E. coli whole cells
with a 0.45 μm microporous (Durapore) membrane.   Inlet pressure
was 90 psi.   The initial volume was 22.3 liters; the final volume,
4.8 liters.

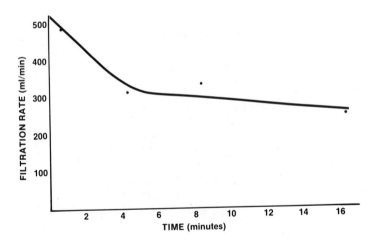

Figure 6.   Flow decay curve for concentrating E. coli lysate with
a 0.45 μm microporous (Durapore) membrane.   Inlet pressure was
90 psi; outlet pressure, 30 psi.   The initial volume was 6.8
liters; the final volume, 1.3 liters.

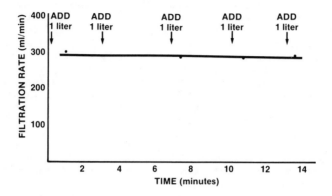

Figure 7. Flow versus time for the lysate wash with 5 square feet of a 0.45 μm microporous (Durapore) membrane. Lysate volume (1.3 liters) remains constant during the wash. Inlet pressure was 90 psi; outlet pressure, 30 psi.

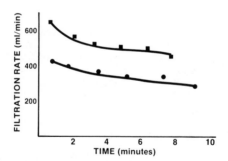

Figure 8. Protein concentration flow decay curves for two different operating conditions with a 100,000 MWCO UF membrane. Key: ■, P_{in} = 80 psi; ●, P_{in} = 40 psi.

circulation rate or higher TMP is responsible for the higher flux.
The absolute flux for the protein solution with the UF membrane is
more than half that seen for concentrating cells with an 0.45 micron
microporous membrane. The protein flux is higher than might be expected, however, the protein fraction has already been clarified
through the 0.45 micron pore size membrane and, hence, is a relatively clean solution.

In an effort to determine whether the trans membrane pressure
(TMP) or recirculation rate are more important in determining the
membrane flux for concentrating cells and lysate or protein, total
recycle experiments were performed. Figure 9 illustrates a typical
response of flux to average TMP for whole cells as circulation rate
is held constant. It is seen that as TMP is increased, the flux
also increases. Figure 10 shows the response of filtration rate to
circulation rate as the average TMP is held constant, again for
whole cells. Beyond a minimum recirculation rate where $P_{in} - P_{out}$
equals about 20 or 30 psi, not much greater flux is generated by
increasing the circulation rate. Cell lysate solutions generated by
sonication exhibit a completely analagous behavior except the absolute flow rates are less than for whole cells. So, for concentration
of either cells or lysate with a microporous membrane, the average
TMP is most dominant in increasing the filtrate rate; the greater
the average TMP, the higher the flux. This should not imply, however, that circulation is not an important variable to manage.

Figure 11 is a flow schematic where the lysate has been produced by a lysozyme digest. The final volume of the lysate was 10.6
liters and a total of 12.7 grams of IgG was assayed in the lysate
suspension. When lysozyme was used to lyse the cells, a very viscous
solution was created due to the large strands of DNA that are generated compared to the sonication method. With sonication, the DNA
tends to be broken up into smaller fragments because of the shear
forces generated. To reduce the viscosity of the lysate suspension,
streptomycin sulfate was added which precipitates the DNA. The
lysate was then concentrated to 700 ml with an 0.45 micron pore size
microporous membrane. After concentration, 6 grams of the IgG remained in the retentate, while only 3.1 grams passed through the
membrane into the filtrate. Concentration of the protein solution
from 9.65 liters to 900 ml with the UF membrane did recover all of
the 3.1 grams of IgG initially in the filtrate from the microporous
membrane.

In an effort to develop a method which would recover more of the
IgG from the lysate concentration retentate, several procedural
changes were made. First, DNAse 1 was added to the lysozyme produced lysate in order to reduce the size of the DNA. Second, a more
extensive washing procedure was instituted. Figure 12 shows the
results obtained with these changes. Initially, there was 3.5 grams
of IgG in the lysate retentate. After a 5 liter constant volume wash
with saline, 2.2 grams of IgG were found in the filtrate, or 61% of
the IgG had been separated from the retained cell debris. Next, 5
liters of saline was added to the 400 ml. During this concentration
step, an additional 0.69 grams (20%) of IgG was recovered in the filtrate. Finally, another 5 volume constant volume wash with 1.5 liters of saline was performed, and 0.25 grams of IgG was flushed
through the membrane. All together, with the DNAse and the extended
washing steps, 88% of the IgG initially in the retentate was re-

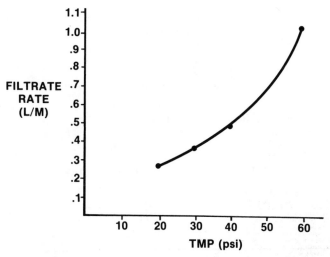

Figure 9. Influence of average transmembrane pressure on membrane flux (0.45 μm microporous). Recirculation rate (40 psi) is kept constant as TMP is varied. Operation is in the total recycle mode.

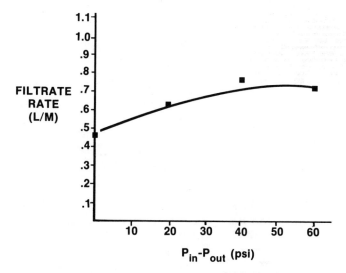

Figure 10. Influence of recirculation rate on membrane flux (0.45 μm microporous). Transmembrane pressure (60 psi) is kept constant. Operation is in the total recycle mode.

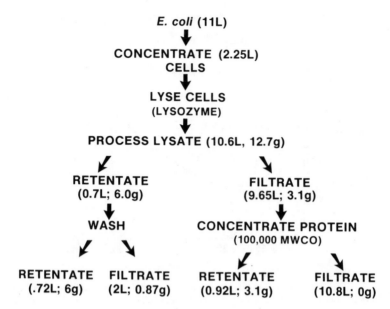

Figure 11. Distribution of IgC for the different process steps. The lysate is generated via the addition of lysozyme to E. coli cells.

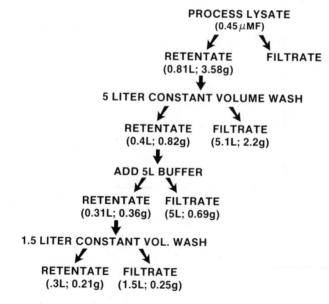

Figure 12. Lysate washing to recover IgC from retentate. Each additional washing of the original retentate will recover more IgC in the filtrate. The first constant volume wash was the most important.

covered. Additional washes would presumably have recovered the last 12% of IgG.

Figures 13 and 14 respectively show the relationship of average trans membrane pressure and circulation rate on flux for both whole cells and lysate suspensions. The lysate was produced via the lysozyme procedure. Qualitatively, the behaviors are quite similar to those seen in Figures 9 and 10 where the cells are lysed by sonication. Beyond minimum circulation flows, little extra flux is attained as the circulation rate increases. The average TMP is the dominant factor in producing flux, however, if P_{in} - P_{out} is less than about 20 psi, debris can accumulate at the membrane surface and inhibit flow. It should be emphasized that these results pertain to 0.45 micron pore size microporous membrane only.

The circulation rate of the protein solution across the top side of the membrane does have an influence on flux with the ultrafiltration membrane. Figure 15 shows a plot of filtrate rate versus circulation rate as measured by (P_{in} - P_{out}) where average TMP is held constant. As the circulation rate is increased so too does membrane flux. The initial increase (P_{in} - P_{out}) has the most profound effect on flux, but subsequent increases in (P_{in} - P_{out}) also gives greater fluxes. Figure 16 shows the results of a total recycle experiment investigating the influence of TMP on membrane flux when lysate protein is processed with a UF membrane. A direct linear relationship is seen. So, to maximize flux for concentrating protein solutions with UF membranes, high TMP's and high circulation rates are needed. This behavior is somewhat different than for cells and lysate with a microporous membrane.

Operating conditions which maximize membrane flux do not necessarily maximize protein throughput. Higher fluxes also transport more material to the membrane surface that can contribute to the polarization layer or can become lodged in the depth of the filter, and these effects will lead the membrane to reject more IgG molecules. To determine which one of several competing effects will dominate to influence the flow of IgG across the membrane, total recycle experiments were performed with the instantaneous concentration of IgG was measured in the filtrate stream as a function of different operational variables. Figure 17 shows both the membrane filtration rate and concentration of IgG in the filtrate versus circulation rate while average TMP is held constant. As the circulation rate and also the filtration rate is increased, so to is the concentration of the IgG in the filtrate at any one time. Figure 18 shows the same effect for increasing average TMP as circulation rate is held constant. Again, as the flux is increased, the concentration of IgG in the filtrate stream goes up accordingly. For the results shown in Figures 17 and 18, the bulk concentration of IgG in the feed is about one gram/liter, and it should be noticed that this bulk concentration does not pass the membrane unabated. The polarization layer formed by the cell debris and fragments is acting to retard the flow of IgG through either the polarization layer or the membrane itself. A solution of IgG alone will be freely permeable through a microporous membrane with an 0.45 micron pore size. However, the concentration values of 70-80% of the bulk feed concentrate can be attained in the filtrate by the proper choice of filtration operating conditions.

As the lysate is concentrated, the instantaneous concentration

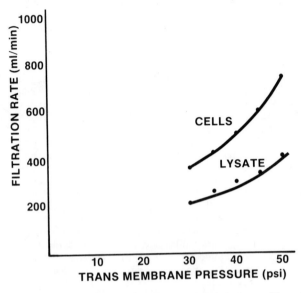

Figure 13. Influence of transmembrane pressure on flux (0.45 μm microporous). Lysate was produced via the lysozyme procedure. The same general behavior as for sonified lysate is seen here. $P_{in} - P_{out} = 30$ psi.

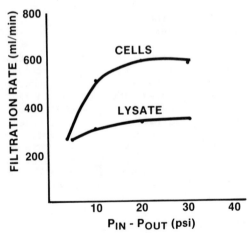

Figure 14. Influence of recirculation rate on membrane flux (0.45 μm microporous). Lysozyme was used to make the lysate. Constant average TMP = 45 psi.

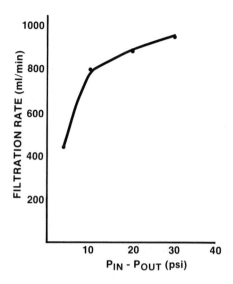

Figure 15.   Influence of the circulation rate on membrane flux (100,000 MWCO UF). At a constant transmembrane pressure (45 psi), the flux of a UF membrane is dependent on circulation rate for a protein solution.

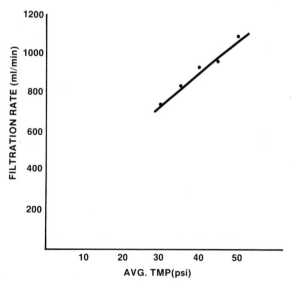

Figure 16.   Influence of transmembrane pressure on membrane flux (100,000 MWCO UF). $P_{in} - P_{out} = 30$ psi.

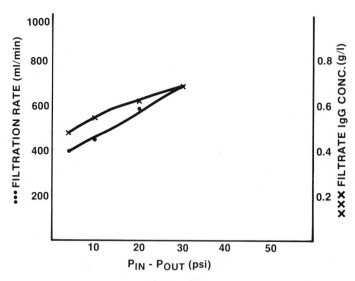

Figure 17. Membrane flux and instantaneous filtrate IgC concentration as a function of filtration rate (0.45 μm microporous).
As the circulation rate or flux increases, the concentration of
IgC in the filtrate increases also. Bulk concentration of IgC
in the feed is about 1 g/liter. Operation is in the total recycle
mode. Constant TMP = 45 psi.

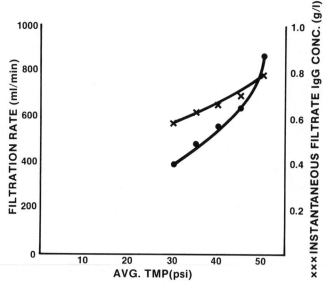

Figure 18. Membrane flux and instantaneous filtrate IgC concentration as a function of transmembrane pressure. Operation is in
the total recycle mode. As flux increases, IgC concentration
increases also. Bulk concentration of IgC in the feed is 1 g/
liter. Constant $P_{in} - P_{out}$ = 30 psi.

of IgG in the filtrate rises. This effect is shown in Figure 19 where the lysate suspension used for the total recycle experiments (Figures 17 and 18) was concentrated. Plotted in Figure 19 are both the membrane filtration rate and concentration of IgG in the filtrate stream as a function of the % conversion of total feed volume to filtrate. The flux remains relatively constant, however, the IgG concentration increases with time or % conversion of feed to filtrate. Theory would predict this effect for a solute molecule where the membrane has a constant rejection coefficient for the protein. Because the IgG molecules in the feed are not freely permeable through the membrane, the IgG will become concentrated with time on the up side of the membrane. For any particular rejection coefficient of the membrane for the IgG, as the IgG concentration increases in the retentate, a higher absolute amount of IgG will pass the membrane. Theoretically for a pure solute solution, the concentration of that solute in the filtrate is described by

$$C_p = C_0 (1-R)(1-\%)^{-R} \qquad (1)$$

where C_p is the instantaneous concentration of solute in the filtrate, C_0 is the bulk concentration of solute on the feed side of the membrane, and % is the % conversion from feed volume to filtrate volume. R is the membrane's rejection coefficient.

Plotting "Equation 1" gives a curve that is qualitatively similar to that seen in Figure 19. It is not productive to fit "Equation 1" to the curve in Figure 19 because in the actual concentration a changing rejection coefficient was observed. From an operational perspective, in order to allow maximum amount of IgG or other protein to pass the membrane, the feed solution should be concentrated to as small a volume as possible. The higher the conversion of feed to filtrate, the greater will be the amount of protein that passes into the filtrate. The last portion of filtrate has the highest concentration of IgG and hence, special care should be taken for the recovery of this portion. On the other hand, if it is desired to retain material, the processed volume should not be concentrated down as far as possible.

Discussion and Conclusions

From the work presented here, it is clear that E. coli lysates can be processed with membrane systems, and that specific non aggregated proteins can be quantitatively recovered without significant losses. With our system, membrane processing starts with cells from a fermentor and ends up with a crudely fractionated and concentrated protein solution. This is all accomplished with basically the same equipment and by using both microporous and ultrafiltration membranes.

In terms of optimizing system performance, the flux for both cells and lysate suspensions seem to be most strongly influenced by the average trans membrane pressure, although maintaining a minimum circulation flow is critical also. Flux rates on microporous membranes for lysates are typically less than for whole cell suspensions as would be expected because of the dispersed cell debris present. Filtrates from lysate processing are typically clear, but do depend on the membrane used and the method of lysing the cells. The ultra-

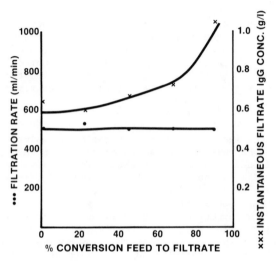

Figure 19. Membrane flux and instantaneous filtrate IgC concentration as a function of conversion of feed to filtrate (0.45
μm microporous). The lysate is being concentrated with time.
The membrane rejection coefficient changes with time, as does the
concentration of IgC in the filtrate. P_{in} = 60 psi; P_{out} = 30 psi.

filtration concentration step proceeds relatively quickly because an already clarified solution is being processed. The flux through the ultrafiltration membrane is more dependent on circulation rate than for the microporous case. For the UF concentration, both high trans membrane pressure and high circulation rates are appropriate. The method of reducing the viscosity due to DNA in the lysate appears to be fairly important. Mild precipitation conditions, without subsequent removal of the DNA will contribute to inhibit the passage of protein through the microporous membrane. DNAse treatment appears to be a better method for removing the DNA step compared to the use of streptomycin sulfate.

Unaggregated protein passage through the microporous membrane has been shown to be flux dependent. If the flow or flux is increased, so too is the instantaneous concentration of the IgG. However, the IgG concentrations in the filtrate, under the conditions tested, were still less than the bulk concentration of IgG in the feed. Under no circumstances did we see passage of the IgG through the 100,000 NMWL UF membrane, and the assay system was sensitive down to 0.01 g/liter. Complete passage of the IgG through the microporous membrane requires some degree of washing. The other soluble molecules or the debris present decrease the effective pore size of the microporous membrane. The amount of washing needed will depend on the method of lysis. Sonication produced lysates resulted in freer passages of IgG compared to lysozyme produced lysates. Although, in both cases, washing was needed for full protein recovery.

Lysate processing with membrane filters is not limited to lab applications(1) as Quirk and Woodrow processed gram and kilogram scale quantities of Pseudomonas sp. for the purpose of isolating enzymes. Compared to centrifugation procedures, the filtration system gave significantly higher specific activities (0.85 vs 0.2) for the enzyme aryl amidase. For carboxypeptidase, upwards of 85% activity was recovered from the original lysate and the filtration system was better in terms of total yield and process time needed compared to centrifugation. These authors concluded that tangential flow filtration possessed a number of advantages, and that filtration was a practical alternative to centrifugation.

Acknowledgments

The authors would like to thank Jack McDowell and Aldo Pitt for their work in performing the IgG assays without which this work would not have been possible.

Literature Cited

1. Quirk, A. and Woodrow, J. R., Biotechnology Letters, 1983 p. 277-282; 5(4).
2. Gabler, R., Developments in Microbiology, (in press)
3. Henry, J. D., "Recent Developments in Separation Science"; 1972 N. N. Li Ed., CRC Press: Cleveland, Vol. 2, p. 205-225.
4. Nelsen, L. and Reti, A. R., Pharmaceutical Technology, 1979 3(5), 51.
5. Sweadner, K. J., Forte, M., and Nelsen, L., Applied and Environmental Microbiology, 1977, 34, 382.

6. Koppensteiner, G. D., Kruger, K., Osmers, W., Pauli, H.,
    Zimmermann, G., Drugs Made in Germany, 1976, 19, 113.
7. Sutherland, I. W. and Wilkinson, J. F., Methods in Microbiology,
    1971, J. R. Norris and D. W. Robbins Editors, Academic Press
    New York, Vol. 5B.
8. Melling, J., Philips, B. W., Handbook of Enzyme Biotechnology,
    1975 John Wiley: New York, A. Wiseman Ed., p. 58.

RECEIVED August 31, 1984

Application, Sterilization, and Decontamination of Ultrafiltration Systems for Large-Scale Production of Biologicals

R. T. RICKETTS, W. B. LEBHERZ III, F. KLEIN[1], MARK E. GUSTAFSON[2], and M. C. FLICKINGER

National Cancer Institute, Frederick Cancer Research Facility, Fermentation Program, Frederick, MD 21701

Several ultrafiltration (UF) membrane systems have been evaluated for large-scale (20-300 liter) recovery of lymphokines, virus and monoclonal antibodies from eukaryotic cells and for concentration of a microbially produced cytotoxic chromoprotein, largomycin F-11. Membrane concentration and recycle of viable eukaryotic cells has been studied in order to produce high concentrations of antibodies and lymphokines. Fetal bovine serum loads (0-15%) have been studied. Some parallel flow UF systems have limitations in flow rate/membrane area, shear forces, membrane reusability (cost), cleanability and potential for scale-up to processing of larger volumes. These applications require the integration of UF into the process without contamination of concentrated supernatant or recycled cells. Hypochlorite, azide and mild base have been found to be effective for chemical membrane sterilization, cleaning and restoration. Effective in situ steam sterilization requires staging. Membrane performance is directly related to preparation, cleaning, and handling procedures.

This brief overview describes some experiences using tangential-flow and dead-end ultrafiltration techniques for concentration of eukaryotic cells, proteins and virus. The data and conclusions presented here have been drawn from process development work employing available apparatus and should be considered preliminary, rather than definitive or exhaustive. Previous ultrafiltration systems have been described (1-14) for both bench and pilot scale separations of proteins and virus. This paper primarily summarizes work on cartridge and sheet filter systems and their application to processes requiring sterilizable and contained systems.

[1]Current address: Cell-Max Corporation, Hagerstown, MD 21740.
[2]Current address: Monsanto Company, St. Louis, MO 63167.

0097-6156/85/0271-0021$08.00/0
© 1985 American Chemical Society

Eukaryotic Cell Concentration and Cell Recycling

Cartridge Systems. Cartridge systems have been used to remove a
variety of particulates from suspensions, but very little has been
done on their applicability to eukaryotic cell culture harvest or
recycling systems. Bench-scale studies were undertaken to explore
readily scalable sterilizable and relatively inexpensive cartridges
for application to eukaryotic cell cultures. Figure 1 shows the
schematic of the first experimental setup for bench-scale dead-end
filtration employing a 2 square foot (sq. ft.) Pall 0.2 micron
pleated cartridge. All culture-contact components were sterilized
by autoclaving. Low positive pressures were used to drive the
filtration. A positive displacement (ColeParmer Masterflex) peristaltic pump was used to control the rate of flow of filtrate. A
pressure differential of 2 to 3 prounds per square inch (psi) was
maintained as well as possible during this 6 liter trial. Maximum
observed pressure differential (delta P) was 4 psi. Culture volume
was reduced to system minimum dead volume after which the pump was
reversed to back-flush loosely adherent cells from the filter
surface. Twenty-five percent of the original cell population was
recovered in the 10X concentrate provided by the back-flushed
material. Figure 2 shows an experimental design which incorporates
a modified filter housing providing a side arm to recirculate cell
culture retentate back to the culture vessel and to sweep cells from
the filter surface. Recirculation was by a double pump head
(CP 7017) at the rate of approximately 1 liter/min. Filtration
rates of 50-200 mls/min. were employed in various tests of this
system. The pressure drop across the membrane did not exceed 2.5 to
3.0 psi. Various specialized baffles were tested in the annulus
between the pleated cartridge and the housing including silicone
jackets and stainless steel spirals, in efforts to increase the
sweeping action of the recirculating culture fluids. Back-flushing
with filtrate was done by reversing the filtrate pump. This alone
did not significantly improve cell recovery. Fresh media entered
the system from either of the supply vessels and was recirculated
following the addition period. This recirculation did improve cell
recovery by 15-25% overall. Cell recoveries in this system were
improved over the dead-end filtration when done by retrograde flow
or back-flush, but only to 40 to 50% of the original cell population
recovered. These techniques indicate that present torturous path
cartridge systems may be quite suitable for eukaryotic cell harvest
and/or harvest followed by semi-continuous growth of such cultures,
as little decrease in recovered cell viability was observed so long
as shear forces were minimized and excessive pressure changes were
avoided. Cell regrowth subsequent to these concentration tests
was superior to the initial growth cycle, or to normal saturation
densities. Low total recoveries indicate that these filters are not
suitable for concentrated cell culture techniques.
    The same cartridge filters with 2-4 sq. ft. surface area were
employed in a pilot-scale production facility for semi-continuous
growth of a murine lymphokine-producing cell line. This was done
as dead-end filtrations, harvesting 80 to 85% of the 20-40 liter
culture as cell-free filtrate and retaining 15 to 20% for inoculum
to be regrown after volume restoration with fresh medium. No
attempts were made to recirculate or back-flush the filter
cartridges.

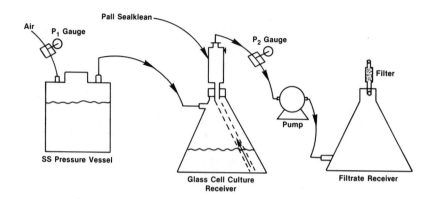

Figure 1.   Cell concentration "dead end" test.

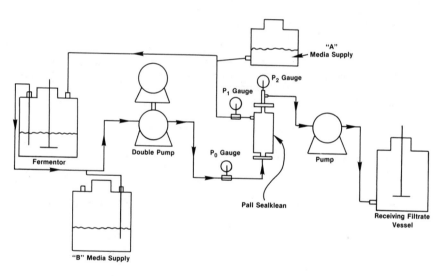

Figure 2.   Cell concentration "recirculation" test.

All components in these product runs including filters were
sterilized in situ by steam. Cooling and drying was accomplished
by sterile air.

Tangential-Flow Ultrafiltration. Filtration systems designed to
incorporate sweeping of the membrane during filtration are less
susceptible to clogging than static systems. Stirred-cell technics
are effective for some macrosolutes but may involve excessive shear
force for eukaryotic cells. Modified plate-and-frame filter systems
incorporating tangential flow and recirculation of retentate should
provide minimum shear while avoiding excessive filter clogging.
Figure 3 shows a cross section of one such system (Millipore
Pellicon). Only retentate flow and manifolds are illustrated. In
this system, pairs of Durapore membranes are separated by mesh
(Figure 4A) or silicone linear channels (4C) with retentate inlet
through the openings at one end of the screen, exit at the opposite
end. In this flow pattern, retentate travels parallel to the
membrane surface, the number of retentate pathways employed varies
with desired filter surface area. Figure 3 represents the Pellicon
unit setup with three membrane pairs, or 1 1/2 sq. ft. of membrane.
Figure 5 shows permeate or filtrate flow paths from the same setup,
with filtrates passing through the membrane and out of the unit
through either or both filtrate manifolds. Figure 6 illustrates
our modification for a serpentine-flow pathway created by blocking
retentate channels in one end of the filtrate screens (Figure 4).
The conventional filtrate screen, (4B), has openings complementary
to those of the retentate screens. Blocking the retentate channels
at one end is performed by silicone (4D). Incorporation of two such
modified screens provides for serpentine-flow through the unit with
the entire retentate volume sweeping each membrane pair. Table I
compares these two techniques.

Table I. Comparison of Parallel and Serpentine Flow

Parallel-Flow Pathway	
Advantage:	Simplicity
Disadvantage:	Doubling filter area halves cross-flow per unit area.

Serpentine-Flow Pathway	
General:	System requires odd number of filter sheet pairs.
Advantage:	Cross-flow per unit area not as rapidly reduced by increased area.
Disadvantage:	Only one left-channel block and one right-channel block allowed per stack, each nearest retentate ports.

This parallel-flow system is that for which the unit was designed,
but for cell harvests or recycling technics it suffers by reducing
filter cross-flow by half for each doubling of filter area - a

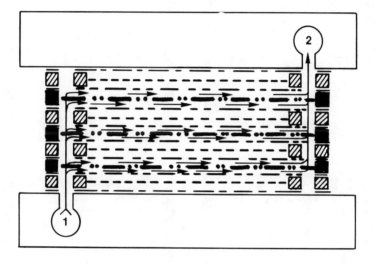

Figure 3. Conventional parallel-flow pathway (retentate flow only shown). Key: 1, retentate inlet distribution manifold; 2, retentate outlet distribution manifold; - — -, end gasket; - - - -, filtrate screen; - - —, membrane filter (Durapore); • • — , retentate screen on linear path retentate channel; ———▶, retentate parallel-flow pathway.

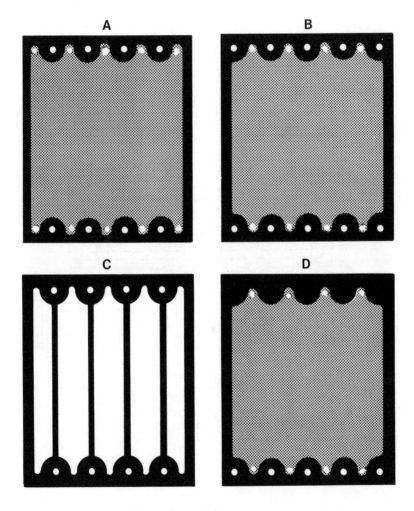

Figure 4.   Membrane screens.

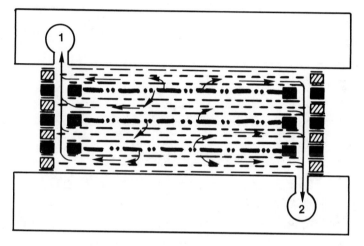

Figure 5.   Filtrate flow paths.   Key: 1 and 2, filtrate manifolds;
– —— –, end gasket; – – – –, filtrate screen; – – ——, membrane
filter (Durapore); ••——, retentate screen on linear path retentate
channel; ——▶, filtrate flow pathway (conventional).

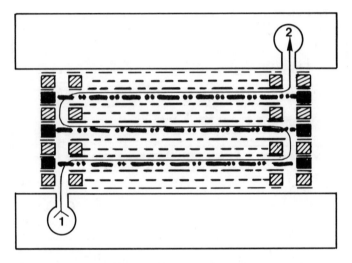

Figure 6. Serpentine-flow pathway generated by modification of
filtrate screens (retentate flow only shown). Key: 1, retentate
inlet distribution manifold; 2, retentate outlet distribution
manifold; – — –, end gasket; – – – –, filtrate screen; – – —,
membrane filter (Durapore); • • —, retentate screen on linear
path retentate channel;——►, retentate serpentine-flow pathway.

situation which has been found to lead to some membrane clogging and
which limits processing volume and area of filter employed. The
1 1/2 sq. ft. unit is the maximum that can be used with parallel
flow and one liter/min. recirculation rate. Exceeding this rate
results in reduced (50%) recovery rather than 90 to 95% cell population recovery after 10-fold concentration. The 1 1/2 sq. ft.
parallel flow system (Figure 7) can process 6 to 10 liters of
approximately 2 million vc/ml cell culture to 10-fold concentration
with recirculation rates of 1 liter/min. and filtration rates of
approximately 100 ml/min. It is possible to shut-off filtration at
approximately 1 liter filtrate intervals or whenever delta P exceeds
2.5-3.0 psi and allow 3 to 5 min. recirculation without filtration
to sweep loosely adherent cells free of the membrane surface.
Filtration is continued to system dead volume, i.e. no culture left
in culture flask, after which the filter is back-flushed with fresh
medium to purge concentrated cell culture. Excessive membrane
clogging can be relieved to some degree by back-flushing with filtrate during recirculation. This may not be necessary where
adequate cross-flow is maintained. The serpentine-flow modification
can be used with 1 1/2 to 4 1/2 sq. ft. of membrane surface, but
does not readily scale beyond this size. The larger surface areas
have not been tested in sterile operation to date. One limiting
factor is the 1/4 inch interior diameter tubing employed throughout
the system. Larger diameter tubing and larger capacity low-shear
pumps, possibly coupled with redrilling of the Pellicon block
retentate manifolds, could allow larger volume processing.

In all eukaryotic cell concentration and recycling procedures,
certain limiting factors have been identified: i) avoidance of
shear by maintaining restriction-free flow and by using minimum
delta P to drive the filtration (for bench-scale cell recycling and
concentration a maximum of 3 to 5 psi inlet pressure was employed,
with filtrate back pressure maintained by filtrate pump rates,
retentate back pressures were 0-2 psi from system resistance);
ii) sweeping the membrane by recirculation with filtrate-flow
stopped to clear loosely adherent cells from the membrane whenever
delta P exceeds 2.5 to 3.0 psi; and iii) backflushing with medium
at the termination of the run to purge the membrane surfaces and
channels of concentrated cell culture.

A similar experimental setup is used for bench-scale testing,
with the Pellicon unit replaced by the Minitan (Millipore) small-
scale serpentine flow system. Figure 8 shows the Minitan filter
packet (right) and the linear channel retentate gasket (left).
Serpentine flow is achieved by alternating left- and right-handed
positions for the retentate path. This unit can hold 1/2 sq. ft.
of filter area in 0.1 sq. ft. packets, and is useful for volumes to
a maximum of 3-5 liters.

Protein Purification

Recovery of Largomycin F-II. Largomycin F-II is a chromoprotein of
approximately 30,000 MW produced by Streptomyces pluricolorescens.
In addition to largomycin F-II, several other biologically active
proteins, ranging in apparent molecular weight from less than 1,000
to greater than 400,000 are present in the fermentation broth.
Using the Biological Induction Assay (BIA) and Micrococcus luteus

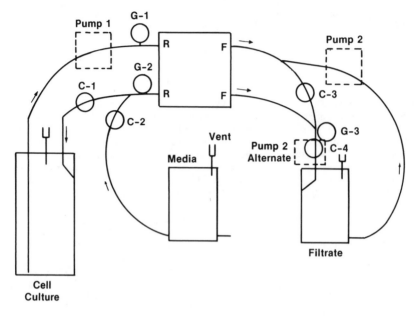

Figure 7. Millipore Pellicon or Minitan operating schematic.

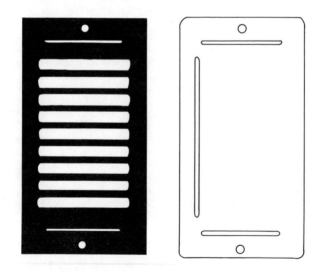

Figure 8. Minitan system.

(ML) as measures for biological activity, these contaminating active
principles must be removed from the largomycin F-II during
processing.

Largomycin F-II may be isolated from S. pluricolorescens
fermentation broth from either the supernatant or the mycelium. The
recovery scheme for the cake purification process and the supernatant recovery process are shown in Figures 9 and 10. The advantages of the cake procedure include low protease activity initially
and higher F-II specific activity. The supernatant contains more
largomycin F-II (up to 400 micrograms/ml of broth) than does a cake
extract, but also contains many more contaminating proteins derived
from the culture medium and consequently requires more processing
steps. Both procedures include several different ultrafiltration
steps.

One step consistent with both processing schemes is the gel
permeation chromatography on Sephadex G-100. Initially, consider-
able difficulty occurred in preparing a concentrated largomycin F-II
rich solution in a sufficiently small volume (2L) having a viscosity
low enough to efficiently and rapidly pass through the G-100 stack
column. By passing largomycin F-II rich materials through a 0.45
micron Durapore filter (Millipore, Pellicon) we were able to remove
enough of the high viscosity materials to produce concentrated
solutions of largomycin F-II (about 20,000 micrograms/ml) which were
ideal for chromatography on the Sephadex stack column. Passage
through the Durapore had the added benefit of removing essentially
all of the F-I fraction from the F-II, eliminating the need for the
ammonium sulfate precipitation step. The flow chart for a typical
experiment is shown in Figure 11. HPLC tracings for each of these
fractions (Figure 12) shows the improved separation with each
ultrafiltration step.

In the course of largomycin F-II process development, the
performance of a variety of ultrafiltration methodology was
evaluated. Millipore, Dorr Oliver and DDS sheet and Amicon hollow
fiber equipment were tested. The data is presented in Table II.
The greater success with the Dorr Oliver system is probably due to
the cellulosic nature of these membranes, as the largomycin F-II
supernatant did contain polypropylene-derived antifoaming agents
which are known to suppress flux rates across polysulfone membranes.
The flux rates for all sheet systems were superior to those observed
with the hollow fiber system.

Recovery of Monoclonal Antibodies. The systems used at the bench-
scale for concentration and partial purification of murine hybridoma
monoclonal antibody were essentially identical to those shown
earlier for membrane (Pellicon) processing of cell cultures.
Instead of the Durapore membranes employed for cell processing,
proteins were concentrated with 100- 30- 10-K dalton nominal
molecular weight polysulfone membranes. The antibody produced is
on the order of 140,000 daltons. Table III emphasizes the impor-
tance of the term "nominal" in molecular weight cutoffs. The
nominal weight is determined by the vendor for standard globular
proteins in specified solutions. When the protein of interest
occurs in small amounts in highly complex protein mixtures such as
tissue culture medium; or when the ionic strength is altered to
influence protein aggregation, and possibly the membrane itself,

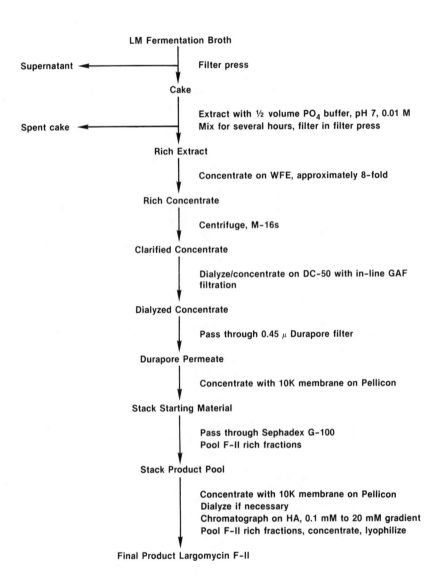

Figure 9. Largomycin F-II cake recovery scheme.

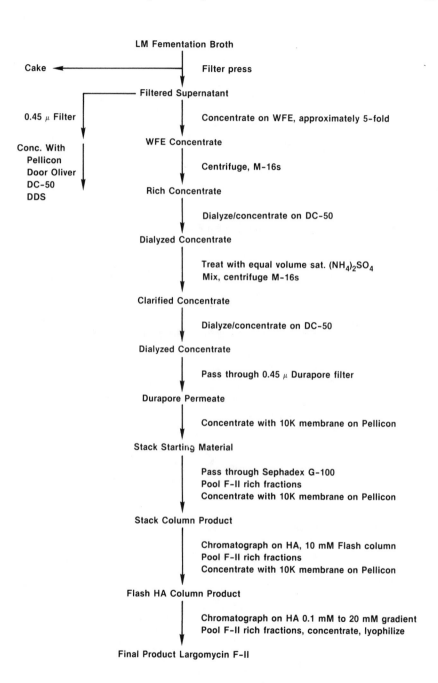

Figure 10.   Largomycin F-II supernatant recovery scheme.

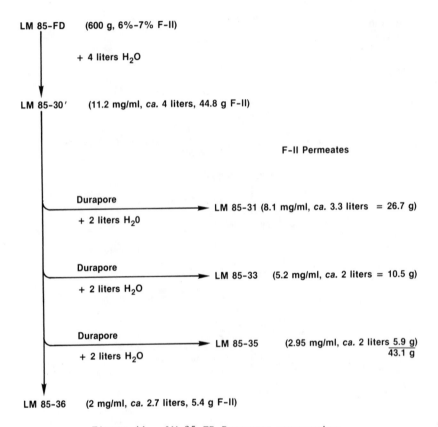

Figure 11.   LM 85-FD Durapore processing.

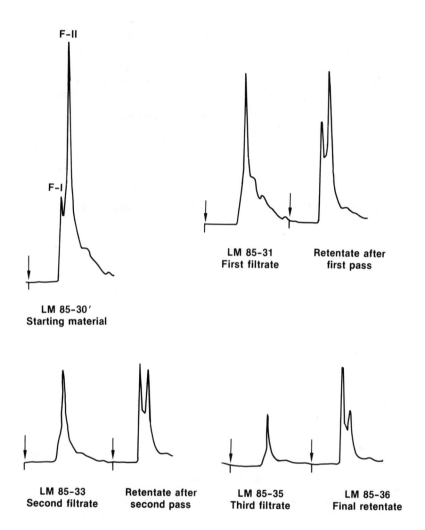

Figure 12. Durapore processing of LM 85 supernatant.

Table II.   Concentration of 0.45 μm Filtered Largomycin F-II
            Supernatant with Ultrafiltration Units

Parameter	Vendor			
	Millipore	Door Oliver	Amicon DC-50	DDS
ft^2	5	1.44	50	2.34
Membrane	10K sheet Polysulfone	10K sheet Cellulose	10K hollow fiber Polysulfone	20K sheet Polysulfone
Volume/ Time	9—2 liter 43 min	10—5 liter 80 min	19—4 liter 30 min	20—12 liter 150 min
GFDs	12.13	50.00	3.87	8.10
PSI I/O	$\dfrac{13}{0}$	$\dfrac{43}{22}$	$\dfrac{25}{15}$	$\dfrac{130}{73}$
Recirculation Rate	3.0 liter/min	11.7 liter/min	24 liter/min	6.5 liter/min

All units operated ≅ steady state with ∼ constant flux rates.

Table III.   D3 Immunoglobin Concentration as Related to Membrane
             Molecular Weight Cutoff

MWCO	0.5 M NaCl	Sample	Titer (mg IgG/ml)	Protein (mg/ml)	Specific Activity (mg IgG/mg protein)	Recovery (%)
100K	+ *	Orig. Susp.	0.065	27	0.0024	80
		Conc.	0.957	107	0.0089	
		Filtrate	0	—		
	+ †	Orig. Susp.	0.024	1.35	0.0178	20
		Conc.	0.137	5.70	0.0240	
		Filtrate	0.016	ND		
	− †	Orig. Susp.	0.086	1.65	0.0515	63
		Conc.	0.726	14.5	0.0372	
		Filtrate	0.005	ND		
	+ †	Orig. Susp.	0.020	1.38	0.0145	36
		Conc.	0.098	4.20	0.0233	
		Filtrate	0.016	1.58	0.0101	
10K	+ †,‡	Orig. Susp.	0.016	0.94	0.0169	>100
		Conc.	0.703	18.25	0.0385	
		Filtrate	0	0	—	

*RPMI-1640 medium + 15% FBS
†RPMI-1640 medium + 15% amniotic fluid + 2% FBS
‡Reprocessed filtrate from 100K

2. RICKETTS ET AL. *Ultrafiltration Systems* 37

nominal molecular weight limit of a given membrane and molecular
weight of the desired protein product are only guidelines. The cell-
free suspensions processed in the first four examples are compari-
sons between a high protein medium and low protein medium. Antibody titers are determined by triplicate samples using an ELISA
assay system. The standard deviation of replicate titers on
concentrated materials is approximately 30%. Cell-free supernatant
titers show standard deviations of approximately 10%. Although
original titers varied, this variation is not related to percent
recovery (data not included on this table). The two significant
variables are protein concentration and ionic strength. In the
first example, approximately 80% of the total proteins present
passed through the 100K dalton membrane, but no antibody could be
detected in the filtrate. Low protein solutions showed significant
losses even though no additional NaCl had been used. When such
solutions were tested with 0.5M NaCl recoveries were unacceptable,
and significant titers of immunoglobulin were seen in the filtrate.
The filtrates from these two examples were pooled and reprocessed
over 10K dalton membrane. Recovery was complete and no activity
could be demonstrated in the filtrate.

Table IV presents data from runs in which cell cultures were
grown in a defined serum-free medium with relatively low total
proteins. Cell-free culture supernatants were processed over a
100K dalton membrane with or without NaCl additions. Both high and
low salt solutions showed some immunoglobulin in the filtrate.
These filtrates were reprocessed over a 10K dalton membrane with
complete recovery of immunoglobulin. Another pair of runs were
processed over a 30K dalton membrane without immunoglobulin appear-
ing in the filtrates, but with relatively high retention of extraneous proteins. The implication from this is that for this particular protein product, crude culture supernatants with protein concentration greater than 3.0 mg/ml will give high recoveries after
salting and processing over a relatively high (100K dalton)
molecular weight membrane with significant removal of extraneous
smaller proteins in the filtrates. The optimal parameters for each
protein must be determined empirically.

Recovery of Lymphokines. A number of pilot-scale production runs
were performed using a murine cell line which is a constitutive
producer of a lymphokine, interleukin-3. This material is approximately 28,000 daltons, and was produced in low protein medium,
1.7% total residual serum. These production runs were processed
using the equipment shown schematically in Figure 13. These runs
may be grouped into small volume 16 to 32 liter batches and larger
volume of 80 to 140 liter production lots. Small volumes were
processed employing the plastic Pellicon cassette with 5 sq. ft. of
10K membrane. Larger volumes used the stainless steel unit with
20 sq. ft. of the same membrane. Figure 14 shows flux curves for
6 small volume and 10 large volume production runs. Plus or
minus two standard deviations are shown for each curve. These give
some measure of the reproducibility offered by this processing
system. Total concentration achieved ranged from 10 to 30 fold
depending upon starting volume. Figure 15 details filtration rates
for the 10 large volume runs, all of which employed the same
membranes. By the third run there was loss in filtration rate, so

Table IV. D3 Immunoglobin Concentration as Related to Membrane
Molecular Weight Cutoff (RPMI-1640 Supplemented Serum-Free Medium)

MWCO	0.5 M NaCl	Sample	Titer (mg IgG/ml)	Protein (mg/ml)	Specific Activity (mg IgG/mg protein)	Recovery (%)
100K	+	Orig. Susp.	0.027	2.2	0.0123	>100
		Conc.	0.446	14.4	0.0310	
		Filtrate	0.005	0.9	0.0056	
10K	+ *	Orig. Susp.	0.008	0.96	0.0083	>100
		Conc.	0.158	11.0	0.0144	
		Filtrate	0	0.02	—	
100K	−	Orig. Susp.	0.027	2.1	0.0130	>100
		Conc.	0.641	12.9	0.0501	
		Filtrate	0.011	0.9	0.0122	
10K	− *	Orig. Susp.	0.009	0.82	0.0198	>100
		Conc.	0.147	9.5	0.0155	
		Filtrate	0	0.02	—	
30K	+	Orig. Susp.	0.055	2.30	0.0239	87
		Conc.	0.514	16.8	0.0306	
		Filtrate	0	0	—	
30K	−	Orig. Susp.	0.011	3.0	0.0038	>100
		Conc.	0.275	12.4	0.0221	
		Filtrate	0	0.1	—	

*Reprocessed filtrate from 100K

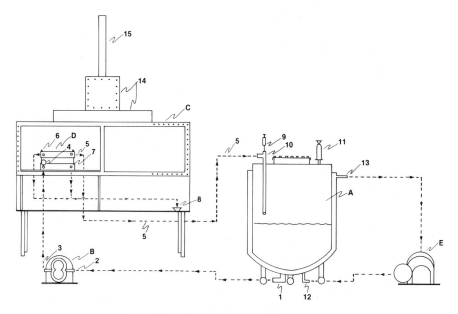

List of Equipment for Batch Ultrafiltration Process

A	Processing Tank Containing Cell-Free Supernatant	Aseptic Sample Port	9	C	Biological Cabinet (negative pressure)	Primary and HEPA Filters	14
		Retentate Return	10			Cabinet Exhaust	15
		Pall Filter (0.1 μm)	11			Contaminate Drain	8
		Processing Tank Outlet	1	D	Millipore Pellicon Cassette System	Product Inlet	4
		Refrigerated Water Jacket Inlet	12			Retentate	5
		Refrigerated Water Jacket Outlet	13			Filtrate Outlets	6, 7
				E	Chiller (compressor)		
B	Tri-Clover Rotary Pump	Pump Outlet Side	3				
		Rotary Pump Inlet	2				

Figure 13. Batch ultrafiltration process.

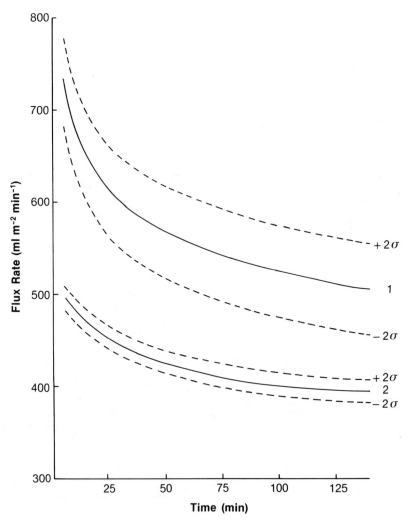

Figure 14. IL3 concentration curves, small and large volume flux
rates. Key: 1, small volume; 2, large volume. (Reproduced
with permission from Ref. 4. Copyright 1984, American Society
for Microbiology.)

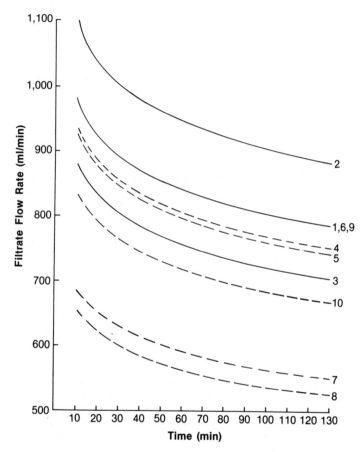

Figure 15. IL3 concentration curves, large volume flux rates, individual runs. Key: ──────, inlet pressure 60 psi; ─ ─ ─ ─, inlet pressure 80 psi.

subsequent runs were performed at increased inlet pressure. Further
variations in the flow rates achieved reflected the care with which
the membranes were cleaned between runs. This set of membranes was
still as useful at the end of this production series as at the
beginning, and thus provides an economical as well as effective
processing step. Recoveries tended to be 80 to more than 100% of
the original lymphokine activity, with no losses seen in the
filtrate.

Virus Concentration

Membrane concentration of viruses has been widely reported in the
literature. Recently this technic has been scaled-up to the
pilot plant scale (Figure 16) for the concentration of greater than
180 liters of virus-rich cell-free supernatant by 15 sq. ft. of
100K membrane. The unit is assembled in a containment room and
sterilized in situ by steam in the configuration shown. The virus
involved is human T-cell lymphoma virus from the C10/MJ-2 cell line,
and may be a significant biological hazard. Figure 17 compares the
concentration process with p24 virus core protein. Flux rates
averaged approximately 1.0 liter/min. using a starting inlet pressure of 50 psi and increasing to 80 psi as flux rate decreased.
Retentate outlet pressures were 10 to 15 psi. Overall concentration was 5.6 fold, while virus core protein (p24) concentration
was 3.8 fold. No core protein activity was seen in the filtrate.
Apparent losses for this preliminary run were approximately 32%.
Whether this loss was due to excessive protease activity generated
during cell removal, from virus adherence to the membrane, or from
excessive shear forces has not been determined. It should be
pointed out that in this preliminary trial, although sweeping was
used, neither back-flushing with filtrate nor back-flushing through
retentate channels was incorporated. These additions to the
technic are under investigation.

Sterilization and Decontamination

Most processing equipment is designed for open operation, so some
modifications of equipment and procedures are commonly required to
employ these systems in biological laboratories where restrictions
apply both to the protection of cultures and products from external
contamination and to the protection of personnel from potentially
hazardous material.

Chemical Decontamination. Standardized procedures for chemical
"sterilization" are generally set forth in manufacturer's
instructions, but such procedures are of limited value when used
in contact with concentrated biological fluids. These procedures,
used with new and clean components, will give a percentage of
sterile runs, but overall are laborious and moderately expensive in
quantity of materials and time required. The present Millipore
Minitan system incorporated steam sterilizable hardware, but the
filter packets themselves cannot withstand steam sterilization.
This system has been tested under chemical decontamination succes-
fully, or it can be aseptically assembled after steam sterilization
of hardware and gas sterilization of filter packets. Neither of

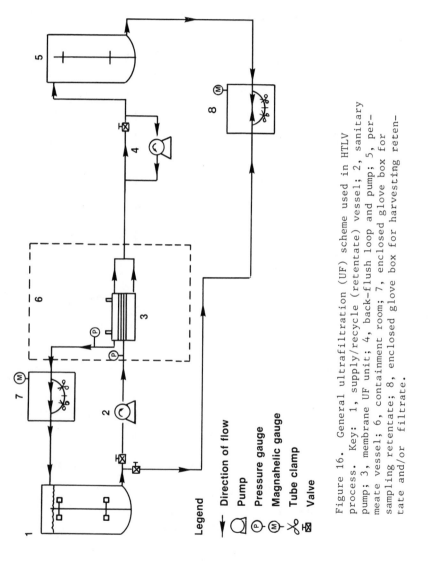

Figure 16.   General ultrafiltration (UF) scheme used in HTLV process.   Key: 1, supply/recycle (retentate) vessel; 2, sanitary pump; 3, membrane UF unit; 4, back-flush loop and pump; 5, permeate vessel; 6, containment room; 7, enclosed glove box for sampling retentate; 8, enclosed glove box for harvesting retentate and/or filtrate.

Legend

→ Direction of flow

Ⓒ Pump

Ⓟ Pressure gauge

Ⓜ Magnahelic gauge

✂ Tube clamp

⊠ Valve

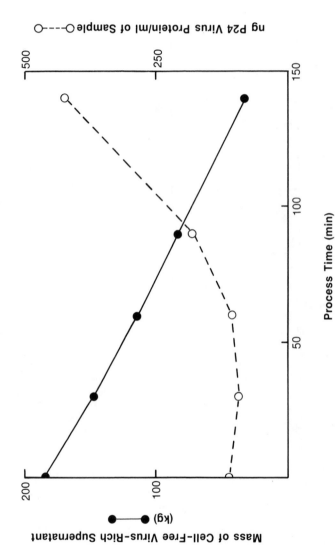

Figure 17. Relationship between volume concentration (formation of virus-rich retentate) and actual virus titer during 100,000 MWCO membrane ultrafiltration (UF). Supernatant was concentrated 5.6 times by this method. Virus was concentrated 3.8 times. The permeate contained no virus activity.

these procedures is of sufficiently high reliability to be employed routinely.

Routine decontamination and cleaning following manufacturer's suggestions, require forethought to insure applicability, safety, and thoroughness. The system decontamination following concentration of the HTLV is a case in point. This agent is not only implicated in human malignancy, but is now under investigation as a possible agent in acquired immune deficiency syndrome (AIDS). Following concentration of HTLV over 100K membrane, the system is processed in situ by the procedure detailed in Table V.

Table V.   Membrane System Decontamination Procedure

In situ, all waste to kill tank:
1. NaOH, 1M, 1 liter/sq. ft. membrane, filtrate valves closed.
2. Repeat step 1 with filtrate valves open.
3. Allowing standing (30 min.) with solution.
4. NaOCL, 525 ppm (1% Chlorox), 1 liter/sq. ft. membrane.
5. Allow standing, overnight, with solution.
6. Flush with R/O water, 205 liters/sq. ft. membrane.

Containment Room
1. Remove filter stack.
2. Place in sodium azide (0.1%) overnight.
3. Store filters at 4°C.

In situ
1. Reseal unit.
2. Steam sterilize system, 30 min. 121°C.
3. Sterilize kill tank.

Low hazard processing systems, such as those discussed for use with monoclonal antibody and lymphokines are also cleaned in situ, but under less rigorous containment conditions, detailed in Table VL Membrane cleaning is essentially that suggested by the vendor, modified in certain applications. Durapore is less chemically resistant than polysulfone, and sodium hydroxide is no longer recommended, but is used as noted in step 3, approximately 1%, without significant damage. This is needed as the Chlorox step is omitted for Largomycin II processes due to the extreme sensitivity of that agent to chlorine.

Flux rates should return to no less than 80% of pre-use values.

Table VI.   Low Hazard Membrane Cleaning

	Polysulfone		Durapore
1.	PBS, 0.5–2.0 liter/sq. ft.	1.	Same.
2.	NaOCl 1%, 1.0 liter/sq. ft., Stand overnight.	2.	NaOCl 0.1%, 0.5 liter/sq. ft. 30 min. to 1 hr.
3.	NaOH, 1M, 30 min.	3.	Not recommended. NaOH, 0.025M, 30 min.
4.	R/O water, 2-5 liter/sq.ft.	4.	Same.
5.	Redetermine flux rate.	5.	Same.
6.	Store wet 4oC	6.	Same.

Steam Sterilization. Cartridge filter systems discussed for cell
harvest applications present no difficulties in steam sterilization.
They may be sterilized by autoclaving, with vents open (and filtered)
for 1 hr. at 121ºC. The same units coupled to fermentation equipment may be sterilized in situ for the same time.

The tangential and serpentine flow membrane filter units tested
are available in non-autoclavable acrylic or sterilizable stainless
steel. As mentioned earlier, the filter packets for the serpentine
flow system will not withstand steam sterilization. The Durapore
and polysulfone filter sheets for the larger units withstand 1 hr.
at 121ºC well, but must be thoroughly wetted prior to sterilization.
Original instructions for separately autoclaving the steel unit
pre-assembled and wetted, suggest "finger-tightening" only of the
retaining nuts prior to sterilization. When this was done in our
laboratories, concentration of eukaryotic cell cultures was much
slower and clogging a much greater problem than had been anticipated.
Examination of the filters themselves showed that significant
shrinkage had occurred, resulting in reduction of retentate and
filtrate manifolds to a fraction of their original size. Poly-
ethylene end gaskets were used in this assembly, and while these
are not recommended for autoclaving, they are apparently sufficiently restrained by the stainless steel blocks that they do not
change dimensions significantly, except for curling beyond the
margins of the filter stack, nor do the filters themselves change
dimension significantly. The screens, however, do shrink significantly carrying the filter sheets, to which they adhere, with them.
This results in some wrinkling of the filter surface, but this does
not seem to adversely affect filtration. The adverse effect comes
from the shrinkage of the stack away from the openings in the end
gaskets. To counteract this, units to be autoclaved separately
should be torqued to the tension appropriate to the membranes
employed. This procedure is comparable to the assembly for in situ
steam sterilization (Figures 18 & 19) where the pre-wetted unit
must be fully tightened to retain steam. Steam lines should have
sufficient upstream filters to protect against loading membrane
surfaces with pyrogenic materials. For in situ sterilization,
temperature-sensitive pencils are again used to insure sterilization
temperature is achieved for adequate time. The autoclaved unit is
best cooled to ambient temperatures in a protective environment
such as a laminar-flow cabinet, and brought back to proper torque
when cooled. In situ sterilized systems, if not contained, should
be re-torqued several times during cooling to insure integrity.

Conclusions

Cartridge and membrane filter systems tested in preliminary studies
in our laboratories have proved to be applicable to eukaryotic cell
culture processes, i.e. cell removal, semi-continuous culture
growth, cell culture concentration and recycling. Present membrane
systems are limited in processing volume; cartridges are scalable,
but have a narrower range of application.

Manipulation of virus, even more than routine cell cultures, is
likely to require stringent containment techniques which will
greatly complicate the apparatus and facilities necessary for processing. Substantial forethought and extensive testing are required
in the design and operation of these processing systems.

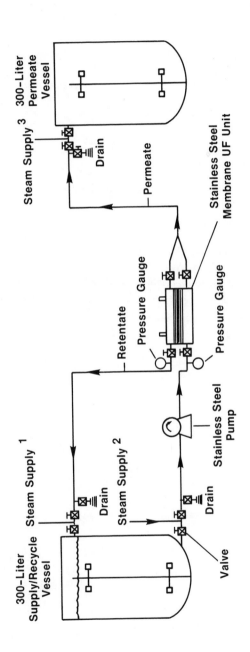

Figure 18. In situ steam sterilization of the membrane ultra-filtration system. (House steam at 22 psig.)

Procedure:

1. Install membranes and tighten UF unit to recommended levels. Open all valves on membrane UF unit.
2. Turn on steam supplies 1 and 2 and open drain at supply 3. Do not open supply 3.
3. Allow 60 min sterilization at 121°C as measured by heat-sensitive pencil.
4. Turn off steam and close drain. Open valve to 300-liter permeate vessel to charge lines with sterile air to maintain positive pressure while cooling.
5. During cool-down, re-torque membrane UF unit to recommended levels.
6. System can be operated when membrane UF unit and pump head are cooled to room temperature.

Figure 19. In situ steam sterilization of the membrane ultra-filtration system.

Most of the technics mentioned here are applicable to sterilization, aseptic processing and post-use decontamination, with similar restrictions to those mentioned for virus and cell culture handling.

Protein processing systems for concentration and/or purification offer a very broad range of application which may be tailored to a given protein product purification scheme. The particular apparatus and operating parameters employed must be selected and refined to the unique requirements of any given protein as well as those of the broth in which it occurs.

Literature Cited

1. Bellini, W.J., Trudgett, A., and McFarlan, D.E. *J. Gen. Virol.* 1979, 43, 633-9.
2. Berman, D., Rohr, M. E., and Safferman, R. S. *Appl. Environ. Microbiol.* 1980, 40, 426-8.
3. Gangemi, J. D., Connell, E. V., Mahlandt, B. G., and Eddy, G. A. *Appl. Environ. Microbiol.* 1977, 34, 330-2.
4. Klein, F., Ricketts, R. T., Rohrer, T. R., Jones, W. I., Clark, P. M., Flickinger, M. C. *Appl. Environ. Microbiol.* 1984, 47, 1023-6.
5. Lee, J. C., Hapel, A. J., and Ihle, J. N. *J. Immunol.* 1982, 128, 2393-8.
6. Mathes, L. E., Yohn, D. S., and Olsen, R. G. *J. Clin. Microbiol* 1977, 5, 372-4.
7. Olsen, A. C. *Proc. Biochem.* 1977, 7, 333-43.
8. Rosenberry, T. L., Chen, J. F., Lee, M. M., Moulton, T. A., and Onigman, P. *J. Biochem. Biophys. Methods.* 1981, 1, 39-48.
9. Sekla, L., Stackin, W., Kay, C., and VanBucken Hout, L. *Can. J. Microbiol.* 1980, 26, 518-23.
10. Shant, J. L., and Webster, D. W. *Proc. Biochem.* 1982, 17, 27-32.
11. Shibley, G.P., Manousos, M., Munch, K., Zelljadt, I., Fisher, L., Mayyasi, S., Harwood, K., Steven, R., and Jensen, K. E. *Appl. Environ. Microbiol.* 1980, 40, 1044-8.
12. Van Reis, R., Stromberg, R. R., Friedman, L. I., Kern, J., and Franke, J. *J. Interferon Res.* 1982, 2, 533-41.
13. Valeri, A., Gazzei, G., Botti, R., Pellezrini, V., Corradeschi, A., and Goldateschi, D. *Microbiology* 1981, 4, 403-12.
14. Zoon, K. C., Smith, M. E., Bridgen, P. A., Zurnedden, D., and Anfinsen, C. B. *Proc. Natl. Acad. Sci. U.S.A.* 1979, 26, 5601-5.

RECEIVED August 31, 1984

Practical Aspects of Tangential Flow Filtration in Cell Separations

JOSEPH ZAHKA and TIMOTHY J. LEAHY

Millipore Corporation, Bedford, MA 01730

Tangential flow filtration is an effective method for performing the separation of cells from a suspending liquid. Cell separation is the unit process of concentrating biomass that has grown during a fermentation. It typically represents the first step in the extraction and purification of product. There are two advantages of using tangential flow filtration over other unit processes to separate cells and products from fermentors. These are the ability: (1) to work in a closed system without generating aerosols; and (2) to effect a more complete removal of the cells from the fermentor effluent. In addition, Tangential Flow Filtration allows for convenient cell washing after concentration in the same system.

The theory of tangential flow filtration as it applies to cell separations is discussed. Major emphasis, however, is placed on presenting the relationship of experimental results to theoretical performance. Topics highlighted are: flux decay with time, effects of operating pressures and flow, membrane fouling, prefiltration requirements and filter geometries.

In most fermentation processes, the fermentation is only the first step in the long process train which includes product formation, recovery, and purification. Commonly the step following fermentation is the separation of the cells from the soluble components in the growth medium. This paper discusses an alternative technique for achieving such a solid/liquid separation.

In the case of extracellular products, emphasis is on the treatment of the product-containing broth. Cells, in this instance, are by-products of the fermentation. The processing of the cells is only important in terms of the potential loss of product with the discarded cells. The primary concern is the clarification of soluble product to increase the efficiency of such operations as product isolation and modification. Intracellular products, on the

0097-6156/85/0271-0051$06.00/0
© 1985 American Chemical Society

other hand, require the concentration of product-containing cells;
the suspending liquid is discarded. Typically, the concentrated
cell mass is then processed to facilitate release of the product.
The desired concentration of cells is often a function of how the
cells will be processed. For example, cell rupture by sonication is
most effective in a narrow cell concentration range.

Several techniques are used to separate cells from the fermentation broth. The most common ones used in large scale fermentations are continuous flow centrifugation, filter presses and rotary
drum vacuum filtration. Tangential (or cross) flow filtration (TFF)
has been proposed as a fourth alternative (2). It is widely used to
process small fermentation batches (<100 liters) and is currently
being evaluated on larger volumes. When compared to other cell
separation techniques, TFF offers some advantages in specific applications. For example, containment of genetically engineered microorganisms is required during their processing of live organisms with
no aerosol formation. Another example involves the processing of
fermentation broths traditionally done by vacuum filtration.
Successful separations using vacuum filtration require the addition
of filtering aids such as diatomaceous earth. These filtering aids
must be separated from the product before its processing and require
disposal after use. Also, there may be significant product loss due
to its absorption to filter aids.

To summarize the principle benefits of TFF, they are:
1. An efficient separation. TFF systems generally have greater
than 99.9% retention of cells.
2. An inherently contained system. A properly designed TFF system
is totally closed to the outside environment. This benefit is important both to contain recombinant organisms and to prevent any
allergenic reactions in workers.
3. Separations independent of the cell/media densities. Separations which are troublesome with a centrifuge due to lack of density
difference between the cells and the media may not be a problem in
TFF.
4. No filter aid addition or disposal.

This paper describes some of our experiences with TFF in cell
separations. We will first present the common components of a TFF
system. This will be followed by some of the performance of TFF.
Finally, we will present several examples where TFF has been used
in industry to process cells.

Tangential Flow Filtration Hardware

Let us examine the TFF technique and a typical process regime.
Tangential Flow Filtration is the general term used to describe
filtration where cross flow parallel to the filter surface is used
to enhance filtration rate. This is in contrast to dead ended
filtration where the fluid path is solely through the filter. If
the membrane used to make the separation is microporous (0.2-0.45 um
pore size), the technique is more specifically called microporous
tangential flow filtration. Ultrafiltration, a subset of Tangential
Flow Filtration, employs a finer, anisotropic membrane able to retain macromolecules, albumin for instance. We will emphasize the
use of ultrafiltration membranes in this discussion.

A semipermeable (ultrafiltration) membrane filter in a tangential flow configuration is used to make a separation based on size. Unlike dead ended filtration, no filter cake builds on the membrane surface during TFF. Rather, the retained species are swept from the surface and mixed back into the bulk solution by the large cross flow parallel to the membrane surface (Figure 1-left). While pressure drop across across the membrane is the driving force pushing filtrate (permeate) through the membrane, the rate of permeation is often dependent on the degree of sweeping action tangential to the upstream side of the membrane.

The retention characteristics of ultrafiltration membranes are measured in Nominal Molecular Weight Cutoff (NMWCO). The NMWCO can be controlled during membrane formation and is typically available in 1,000; 10,000; 100,000; and 1,000,000 daltons. For cell harvesting, ultrafiltration membranes of 100,000 NMWCO and microporous membranes are generally employed. While the separation is based on size, the efficiency of the operation can be influenced by other factors.

In its simplist form, the major hardware components of the ultrafiltration system are the storage tank, pump, and membrane package (Figure 2). Most of the output from the pump sweeps the membrane tangentially and returns to the process tank. The permeate flow is generally less than 10% of the total flow to the membrane.

In the case of extracellular products which pass through the membrane, the permeate is collected for further processing. The cells are concentrated, reducing the initial volume 10 to 20 times. Product yields can be increased by a process called constant volume wash. (Water or buffer is added to the concentrated cells by this technique and permeated out with additional product while maintaining a constant volume of cells.) The TFF operating techniques for intracellular products are essentially the same as for extracellular products but the purpose is different. Permeate is removed to obtain a high concentration cell suspension. Constant volume washing can be employed to remove low molecular weight media components or cell by-products which, in this case, contaminate the concentrated cells.

Figure 3 shows a more detailed schematic of a TFF System in a cell harvesting application. In this particular set up, the system is designed for extracellular product processing but the principal components would also be used for intracellular products. It shows the plumbing of a clean-in-place system and a tank for collecting waste. The system is closed, in that all waste solutions existing the system enter a kill tank for sterilization.

A prefiltration step may be required before processing through the membrane. Many media components used in the production of antibiotics are insoluble. For example, soy grits and calcium carbonate which are not fully utilized during fermentation may be present in the broth at the time of harvesting. They are in the form of large particles and agglomerates. In the case of narrow flow channel TFF devices such as spirals, hollow fibers and some plate and frame devices, these large particles must be removed to assure that the flow channels of the TFF system remain open. The type of prefiltration employed depends on the volume of solution and the mass of contaminant to be removed. Inline strainers, bag filters, or vibrating screen baskets can be used. In the case of fermentations with single celled organisms grown in true solutions, prefiltration to remove particulates is generally not required. An inline strainer is,

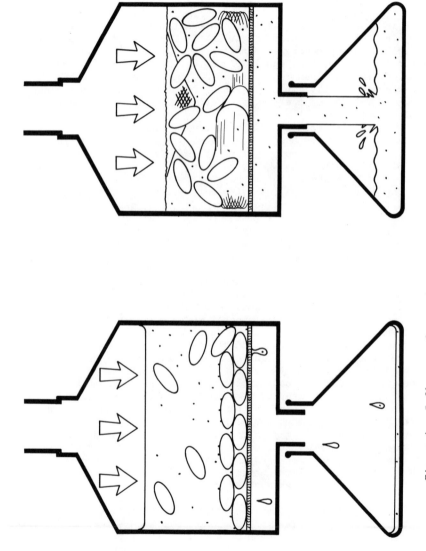

Figure 1. Influence of removing retained species from membrane surface.

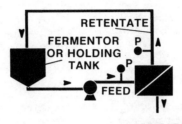

Figure 2. Simplified schematic of tangential flow filtration system.

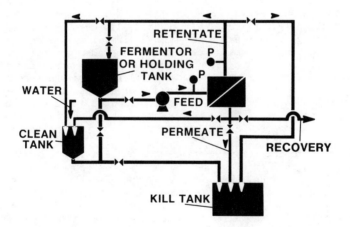

Figure 3. Simplified cell separation schematic of tangential flow filtration system.

however, still desirable to protect the pump and modules from an oc-
casional contaminant.

Performance Of Ultrafiltration In Cell Harvesting

Once it is determined that TFF offers benefits as a separation technique and the separation has been shown to be feasible on the small
scale, the optimization of the system for scale-up begins. The performance parameter which is commonly optimized during scale up is
filtration rate (flux). The flux of the TFF system is dependent on
many variables. In this discussion we will show some typical relationships of flux to these variables.

Pressure and Flow. Pressure and flow to the TFF module are the
easiest operating conditions to control. Based on experience with
dead-ended filtration, it may seem logical that increasing pressure
will increase membrane output. This is certainly the case when we
filter clean water. However, where there are species that are retained on the membrane, tangential flow is required to sweep the surface clean. Often increasing pressure alone will not increase filtration rates. In processes whose filtration rate is limited by the
layer of retained species (gel layer), membrane output is theoretically related to flow to the 1/3 power in the laminar regime and not
related to pressure above a minimum level (1). On the other hand,
Henry and Alred (2) predict that because cells are large in size and
the resulting gel layer has high permeability, flux would be relatively independent of cross flow rate and dependent on pressure.
    Our experiments have shown effects of both pressure and flow on
filtration rate. Figure 4 is a plot of flux versus pressure for an
E. coli in AZ broth. Increasing pressure resulted in increasing permeate output. In addition, flux is also related to flow to about 1/3
power. The actual relationships of flux with pressure and flow, however, vary from one system to another and are dependent on the solution consistancy and fouling (which will be discussed).
    The relationship of wall shear rate to flux for two membrane de-
vices is shown in Figure 5. Wall shear rate is proportional to velocity divided by channel height. These experiments were run with E.
coli grown in a defined salts medium. The average transmembrane
pressure was not held constant but allowed to rise as a natural con-
sequence of the increasing flow. The net result was a dependence of
flux to shear rate to the 1/2 power.
    In general, it is desirable to run the device at as high a flow
rate as possible (determined by the physical characteristics of module). The outlet pressure of the module should be at least 10 psig
to assure sufficient driving force on the low pressure end of the
membrane.
Temperature. Flux is also influenced by temperature. As the temperature of the broth increases, viscosity decreases resulting in
higher flux. There is, of course, practical limits of increasing the
temperature due to the heat labile nature of biological products.
Also, temperatures high enough to result in cell rupture would obviously cause flux decay due to the release of debris and high molecular weight materials from the cells into the fluid.
Concentration. Increasing concentrations of cells causes a decrease
in flux. Theory predicts when filtration rate is controlled by the

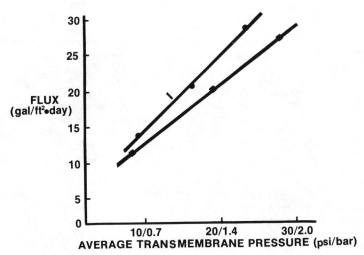

Figure 4. Flux versus transmembrane pressure and recirculation rate for an E. coli in AZ broth. Key: ●—●, 4 gpm (15.2 lpm); ■—■, 2 gpm (7.6 lpm). Membrane used: 15 ft^2 (1.4 m^2) 100,000 MWCO. Note: 10 gal/ft^2 x day = 17 l/m^2 x h.

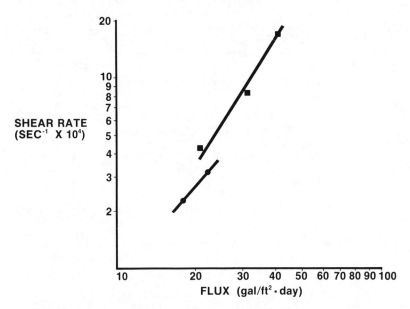

Figure 5. Flux versus membrane shear rate for an E. coli in a defined medium. Key: ■—■, device B (5 ft^2/0.45 m^2 cassette); ●—●, device A (15 ft^2/1.4 m^2 spiral). Membrane used: 100,000 MWCO. Note: 10 gal/ft^2 x day = 17 l/m^2 x h.

start retained species that form a layer on the membrane, the flux
will be inversely related to the logarithm of concentration (1).
Such an effect is shown in Figure 6. This relationship only holds
true under constant flow and pressure conditions. Under normal
circumstances the broth becomes more viscous as the concentration of
cells increases. For example, the cell mass from the experiment depicted in Figure 6 became so viscous at 8.5 fold concentration that
the cell mass in a beaker would not flow out when the beaker was in-
verted. Due to physical limitations of the pump and membrane device
constant cross flow could not be maintained during this concentration. Tangential flow decreased as concentration increased. In
this case the decrease of flux with increasing concentration was due
to both the increased concentration and the reduced tangential flow.
Media Components. Components of the medium, the cells and cellular
products can also affect the filtration rate. This can happen in
two ways, direct effects and membrane fouling. Since membrane
fouling is the major performance limitor in the routine practice of
TFF, we will treat it separately. Direct effects are those which
influence either the sweeping action across the membrane or the
effective permeability of the membrane. Obviously low molecular
weight water-soluble components, such as sugars, which increase the
viscosity of the permeate, will have a direct effect on flux. High
molecular weight components which increase viscosity of the feed,
but not the permeate (polysaccharides, for instance), will cause a
decrease in flux due to reduced shear at the membrane surface (as a
result of pumping limitations), resulting in decreased back diffusion. Another type of direct effect is caused by a component
which is retained by the membrane and becomes involved in the gel
layer of the membrane. These components generally reduce the flux
but enhance the influence of tangential flow rate on output.
Fouling. The most critical relationship to monitor during the
scaling from the laboratory to production is the influence of time
on flux. For some systems, when flow, pressure, temperature and concentration are maintained constant over time flux will also be stable
(Figure 7). However, often the flux will decrease rapidly with time
even when other variables are held constant. In Figure 8 we see a
decrease of flux by a factor of three over a 60 minute period. This
loss in output is called membrane fouling. Understanding its causes
and controlling its effects are the primary responsibilities of the
process engineer when scaling up TFF systems.

Howell and Velicangil (3) described three phases in flux loss
with time. The gel layer of retained species forms on the membrane
in seconds and, as discussed earlier, its restriction on filtration
rate can be reduced by increasing the cross flow. Over a period of
minutes adsorption of constituents from the media on the membrane
takes place. In the time frame of hours, the gel layer on the membrane may become unstable resulting in a less permeable layer. These
effects of adsorption and gel layer instability are the principle
causes of fouling. They result in lower system output than would be
expected based on the solution and operating conditions. The filtration rate of a badly fouled system is dependent on pressure and independent of cross flow.

Even when fouling occurs, steps may be taken to compensate for
or control its effect on system performance. Control of fouling is
central to providing an efficient TFF system and an economical

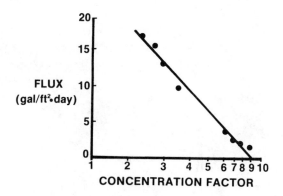

Figure 6. Flux versus organism concentration for a B. Thuringiensis. Membrane used: 15 ft^2 (1.4 m^2) 100,000 MWCO. P_{in}/P_{out}: 15-45/12-8 psi (1-3/0.8-0.5 bar). Note: 10 gal/ft^2 x day = 17 l/m^2 x h.

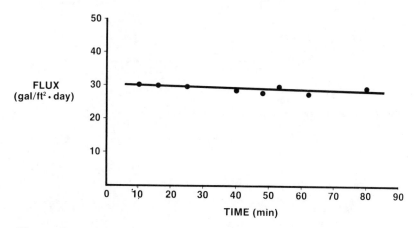

Figure 7. Example of a nonfouling system for an E. coli in AZ broth. Membrane used: 15 ft^2 (1.4 m^2) 100,000 MWCO. P_{in}/P_{out}: 27/21 psi (1.8/1.4 bar). Note: 10 gal/ft^2 x day = 17 l/m^2 x h.

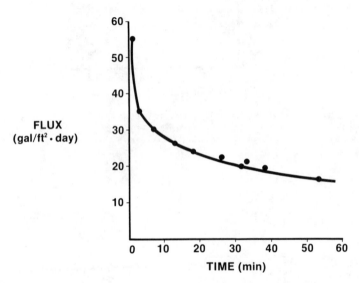

Figure 8. Example of a fouling system for an E. coli in a
defined medium. Membrane used: 15 ft² (1.4 m²) 100,000 MWCO.
P_{in}/P_{out}: 36/23 psi (2.4/1.6 bar). Note: 10 gal/ft² x day =
17 l/m² x h.

separation process. Because fouling is important we have developed
several ways to reduce its effect. In general, combinations of anti-
fouling techniques are used concurrently.

The most straight-forward control of fouling comes with the
choice of membrane. Figure 9 shows the dramatic effect of membrane
type on fouling. Both the 10,000 and 100,000 NMWCO ultrafiltration
membranes are of the same polymer yet the 10,000 NMWCO fouls almost
instantaneously. If the fouling is caused by the adsorption of a
component from the medium, the membrane with the smaller pores would
have greatest reduction in effective pore size. (The more open
microporous membrane of a different polymer material shows an
initially higher flux than the 100,000 NMWCO membrane but virtually
the same flux after 30 minutes.)

Effects such as those shown in Figure 9 may also be due to a
component retained by mechanisms other than adsorption with the
10,000 NMWCO membrane but passing through the 100,000 NMWCO membrane.
This effect can be verified by evaluating the effect of cross flow on
flux. If adsorption is the cause of lower output then increased
cross flow will not increase filtration rate. The opposite would be
true if lower fluxes were due to material retained on the membrane
by size exclusion. Adsorption can also be determined by analyzing
the composition of the feed and permeate to see if there is a difference.

The composition of the membrane can play a role in adsorption.
It may be possible to modify the membrane in such a way as to reduce
or eliminate the binding of a fouling agent. These modifications
can be tested empirically or screened theoretically if the nature of
the foulant is known.

The formulation of the medium which contains fouling agents can
also be modified to reduce membrane fouling. Often, alternative nonfouling components may be substituted or the concentration of fouling
components reduced to increase filtration rate. Antifoams are
examples of a medium component which can cause fouling. Figures 10
and 11 illustrate the effects of antifoams on filtration rate and
some techniques for minimizing output loss. First, the type of antifoam can influence fouling. As shown in Figure 10, the flux loss of
a broth with Antifoam A is not much more than a broth with no antifoam. Antifoam B, however, causes a more severe flux decline. The
concentration of antifoam also affect output. Figure 11 shows the
dramatic effect of antifoam concentration on flux decay. Increasing
concentrations of antifoam increase the loss of filtration rate.
One approach to control the concentration of antifoam is to use the
minimum amount needed to stop foaming. This may be done with an
automatic antifoam addition system. Excess antifoam which is not
associated with the cells or media constitutents is more likely to
adsorb on the membrane, so better control of its concentration will
lower antiforam adsorption.

Techniques also exist for dealing with the gradual instability
of the retained gel layer on the membrane. These techniques are
generally mechanical. For example, Figure 12 demonstrates the benefit of periodically closing the permeate valve. With the permeate
valve closed, the membrane sweeping action is enhanced. In addition,
there is some backflow through the membrane loosening any destabil-
ized layer.

Cleaning is a fouling control technique effective with both

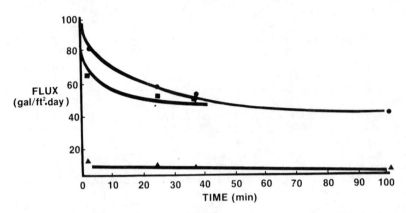

Figure 9.   Effect of membrane type on flux for an E. coli in L
broth.   Key: ●, 0.2 μm; ■, 100,000 MWCO; ▲, 10,000 MWCO.
P_{in}/P_{out}:   30/0 psi (2/0 bar).   Note: 10 gal/ft^2 x day = 17 l/m^2
x h.

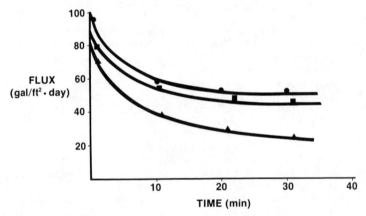

Figure 10.   Effect of antifoam type on flux loss of AZ broth.
Key: ●, no antifoam; ■, antifoam A; ▲, antifoam B.   Note:
10 gal/ft^2 x day = 17 l/m^2 x h.

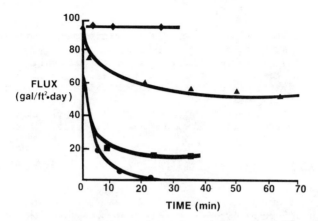

Figure 11.   Effect of antifoam concentration on flux for an E.
coli in a defined medium.   Key: ♦, clean water; ▲, 0 ml/l;
■, 0.1 ml/l; ●, 0.5 ml/l.   Membrane used: 0.5 ft^2 (0.045 m^2)
100,000 MWCO. P_{in}/P_{out}: 28/5 psi (1.9/0.3 bar). Note:
10 gal/ft^2 x day = 17 l/m^2 x h.

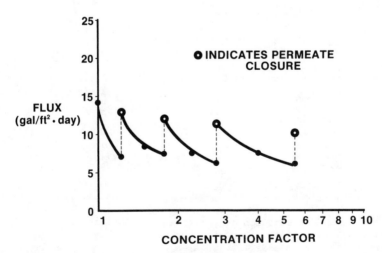

Figure 12.   Effect of permeate closure on flux for Streptomyces
in a complex medium.   Note: 10 gal/ft^2 x day = 17 l/m^2 x h.

adsorption and gel layer instability. If system and batch sizes
allow for relatively short runs, it may be the best means for
dealing with fouling. In other words, you let fouling occur to some
acceptable level during processing and then remove the foulant between processing runs. In Figure 13 we can see that, although the
flux decays with time, the cleaning was effective in bringing the
initial flux back to the same level for successive runs. Choice of
cleaning chemicals is based on the nature of the foulant. Several
types of solutions may be required to both clean and sanitize the
membrane system. The chemicals are generally evaluated empirically
after a preliminary screening based on chemistry and experience.

The information presented here may appear complex, but it is
based on experience gained in many cell separation applications.
Understanding the trends in performance has greatly simplified the
process.

Applications Of Tangential Flow Filtration In Cell Separation

Currently, there are many examples of cell processing in the industrial environment using tangential flow filtration. To illustrate
the breadth of microbial types which may be processed by this technology, we will discuss three applications which have been in routine operation under production conditions. The applications include
cell/growth medium separations directly from fermentors (Escherichia
coli and Mycoplasma species) and the concentration/washing of influenza virus used in the production of flu vaccines.

Table I contains the operational parameters employed during the
harvesting E. coli from a production fermentor.

Table I. E. coli Harvesting Operational Data

Batch Size	400 liters
Area	$25 \ ft^2 (2.3m^2)$
Membrane Used	100,000 MWCO
P_{in}/P_{out}	40/25 psi (2.7/1.7 bar)
Recirculation Rate	3.5 gpm (800 lph)

This particular fermentation is performed in 400 liter batches. Tangential flow filtration was carried out with a Pellicon cassette
(Millipore Corp.). The filtration area was 25 ft^2 using a 100,000
molecular weight cut off ultrafiltration membrane. The inlet pressure was maintained at 40 psig while the outlet pressure was 25 psig
and recirculation rates of the cell suspension were 3.5 GPM tangential to the upstream membrane surface. Pumping of the cell suspension was performed with a Moyno screw pump.

Table II outlines the performance of the filtration system in
this application.

Table II. E. coli Harvesting Performance Data

Initial Volume	400 liters
Final Volume	20 liters
Concentration Factor	20X
Average Flux	15 GFD (26 1/h·m^2)
Recovery Initial Titer	1X10^9
Final Titer	2X10^{10}
Processing Time	6 hours

A 20X concentration was obtained in a six hour processing time with an average flux of 15 GFD. Titrations of the initial and final viable cell counts illustrate the efficient recovery of the cells in this process with no loss of viability.

Mycoplasma are fermented for the production of veterinary vaccines and diagnostic tests. In this particular fermentation, the cultivation medium, PPLO broth, contains serum as a growth component and the fermentation is typically carried out in 250 liter volumes. Successful separations are performed with 50 ft^2 of a 100,000 NMWCO ultrafilter in the same configuration and under the same operating conditions as the E. coli example (Table III).

Table III. Mycoplasma Harvesting (Veterinary Vaccine) Performance Data

Batch Size	250 liters
Growth Medium	PPLO Medium (Serum)
Area	50ft^2 (4.5m^2)
Membrane Used	100,000 MWCO
P_{in}/P_{out}	40/25 psl (2.7/1.7 bar)
Recirculation Rate	3.5 gpm (800 lph)

Concentration factors of 50X (250 liters to 5 liters) were obtained in 6 hours yielding an average flux of 5 GFD (Table IV).

Table IV. Mycoplasma Harvesting (Veterinary Vaccine) Performance Data

Initial Volume	250 liters
Final Volume	5 liters
Concentration Factor	50X
Average Flux	5 GFD $(8.9\ 1/h \cdot m^2)$
Recovery Initial Titer Final Titer	1.5×10^8 3.5×10
Processing Time	6 hours

The lower flux associated with this application compared to E. coli harvesting reflects the increased viscosity of the growth medium due to serum and a slight influence of lower cross flow per square foot. Recovery data indicates some loss of cell viability (23X concentration factor) during processing but it is not surprising considering the fragile nature of the organism.

Both E. coli and the Mycoplasma harvesting applications illustrate some benefits realized by tangential flow filtration users. First, such systems provide efficient containment of the organisms during processing. Thus, the risk to production workers from organism-containing aerosols is minimized. Second, harvesting applications at these volumes are easily performed with small, portable devices which can be transported from fermentor to fermentor as the need arises. Third, routine cleaning and sanitization procedures eliminate cross contamination when the same system is used to harvest different fermentations.

The last cell processing example highlights the utility of TFF as an adjunct to other separation techniques. The production of flu vaccines is a multistepped process involving virus cultivation in eggs; clarification of allantoic fluid; and, virus inactivation, concentration and washing prior to sterile filtration. Classical techniques for concentration and washing use zonal gradient centrifugation followed by dialysis of the virus containing fraction. Such operations require multiple ultracentrifuges for on the order of 8 hours to process a typical batch. These centrifuges represent high capital, maintenance and labor costs. Dialysis, using conventional tubing, is an additional 24 hour processing step which removes low molecular weight components from the vaccine preparation (Figure 14).

Ultrafiltration was used to speed up the concentration step and as a substitute for dialysis in washing the concentrated virus. In this modified centrifugation process, the total number of centrifuges was reduced almost in half and the processing time with each centrifuge lowered from 8 hours to 4 hours. Final concentration of a 30 liter batch was accomplished with 30 ft^2 of 100,000 NMWCO ultrafiltration membrane. Inlet pressure was 20 psi and the outlet pressure was 10 psi (Table V).

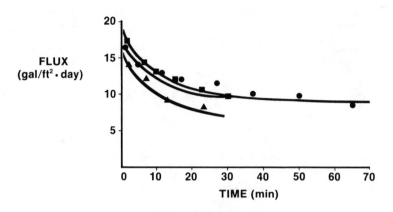

Figure 13. Effect of cleaning on flux for an E. coli in a defined medium. Key: ■, first run; ▲, second run; ●, third run. Membrane used: 15 ft^2 (1.4 m^2) 100,000 MWCO. P_{in}/P_{out}: 20/8 psi (1.4/0.5 bar). Note: 10 gal/ft^2 x day = 17 l/m^2 x h.

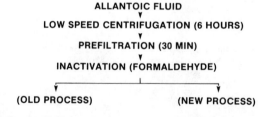

Figure 14. Influenza virus processing schematic.

Table VI contains the performance data using tangential flow
filtration for concentration and washing of influenza virus.

Table V.   Influenza Virus Concentration (Whole Virus Vaccine)
Performance Data

Batch Size	30 liters
Area	30 ft^2(2.7m^2)
Membrane Used	100,000 MWCO
P_{in}/P_{out}	20/10 psl (1.4/0.7 bar)

Table VI.   Influenza Virus Concentration (Whole Virus Vaccine)
Performance Data

Initial Volume	30 liters
Final Volume	2 liters
Concentration Factor	15X
Recovery UF	70-90%
Centrifuge/Dialysis	40-50%
Process Time UF	14 hours
Centrifuge/Dialysis	38 hours

A total of 30 liters were reduced to 2 liters (15X). The two liters
were washed of low molecular weight components by constant volume
washing using the same filtration device. Including the partial processing time of the centrifuges, the entire concentration /washing
procedure took 14 hours with tangential flow filtration compared to
38 hours using centrifugation/dialysis. Recovery data indicates an
additional advantage to the filtration method. Final virus recovery
was consistently higher by filtration than centrifugation/dialysis
(70-90% versus 40-50%). There are less product losses during dialysis since filtration eliminated bag breakage and lower processing
costs because centrifuges were used for shorter times.
All three of these applications are examples of commercial processes using tangential flow filtration. Clearly, the selection of

membrane, filtration area and processing parameters is a function of the individual application and the objectives of the separation. Nonetheless, these applications show that tangential flow filtration is an alternative separation techniques which offers distint advantages over other methods in certain circumstances.

Summary

It was the object of this presentation to outline some of the key parameters which are evaluated during the application of tangential flow filtration for cell separations. A throrough understanding of what is required from the separation is an essential first step. This is followed by laboratory and pilot scale trials where the variables which affect performance are determined. Finally, a system which meets the requirements of the separation is designed and the operational conditions are established for routine use. This was the process followed by these authors when installing the three filtration system discussed earlier.

Literature Cited

1. Blatt, W.F. et al.; "Solute Polarization and Cake Formation in Membran Ultrafiltration: Causes Consequences and Control Techniques," Ultrafiltration Membranes and Applications, Polymer Science and Technology, Volume 13.
2. Henry, J.D. and Alfred, R.C.; "Concentration of Bacterial Cells by Cross Flow Filtration," Dev. Ind. Microbiol., 13, 177 (1972).
3. Howell, J.A. and Velicangil, O.; "Theoretical Considerations of Membrane Fouling and Its Treatment with Immobilized Enzymes for Protein Ultrafiltration, "Journal of Applied Polymer Science, Vol. 27, 21-32 (1982).

RECEIVED September 12, 1984

Preparative Reversed Phase High Performance Liquid Chromatography
A Recovery and Purification Process for Nonextractable Polar Antibiotics

R. D. SITRIN, G. CHAN, P. DePHILLIPS, J. DINGERDISSEN, J. VALENTA, and K. SNADER

Smith Kline & French Laboratories, Philadelphia, PA 19101

Reversed phase high performance liquid chromatography (RPHPLC) has found wide use as an analytical tool in monitoring antibiotic fermentation production, isolation, and purification schemes. The recent introduction of suitable instrumentation and affordable preparative reversed phase packing materials allows the traditional advantages of RPHPLC – speed and resolution – to be applied to preparative work. Classically, isolation and purification processes for polar, charged, non-solvent extractable antibiotics such as peptides or glycopeptides have required multistep, medium to low resolution procedures such as charcoal, cellulose, ion-exchange or size exclusion techniques often with low recovery. However, by the nature of its mechanism, reversed phase HPLC is particularly suited to solve these separation problems because undesirable very polar highly colored contaminants elute at the solvent front ahead of the desired material. Thus, it is possible to introduce such a step very early in a purification scheme on a relatively crude isolate. We will demonstrate examples of such one step purifications on very crude fermentation isolates to yield highly purified homogeneous products from milligram to gram and potentially kilogram scales.

One problem frequently encountered in the recovery and purification of fermentation products such as antibiotics and peptides is the necessity for multistep procedures to obtain pure materials. For substances which cannot be extracted into ethyl acetate or methylene chloride and which typically cannot be chromatographed on silica, purification often requires repetitive chromatographic

0097–6156/85/0271–0071$06.00/0

steps on adsorption media such as polyamide, Sephadex, Biogel,
charcoal and ion exchange or XAD resins. Such multistep schemes
often yield products in low recovery and, being limited in
capacity, are difficult to scale-up to process level. However,
reversed phase high performance liquid chromatography (RPHPLC),
widely used as an analytical procedure (1,2), has great utility as
well in the preparative mode for isolation and purification, often
in one step.

A recent Chemical Abstracts search of the literature (1980-
1983) indicated that, out of 7,000 postings for HPLC papers, only
100 discussed preparative work. Furthermore, the bulk of the
preparative papers were either theoretical discussions (3-8),
descriptions of large scale (multigram) separations on silica or
preparative reversed phase work on analytical columns and instruments (100 mg scale) (9-14). The most recent development in this
area is displacement chromatography (15-18), but so far this has
been applied only in limited cases on analytical scale instruments.
Although several excellent reviews on preparative HPLC have been
published (19-21), few papers describing gram or larger preparative reversed phase chromatography were evident (22-23). Several
successful examples of such separations are described here, ranging from 100 mg to gram scale. Furthermore, these separations
also demonstrate efficient one-step procedures for preparing pure
products starting from crude fermentation isolates. The recent
introduction of industrial scale equipment and packings from
Waters, Whatman and Elf-Aquitaine allows these procedures to be
scaled up to process level.

Background

RPHPLC is performed on columns packed with silica gel to which a
hydrocarbon, usually with 8 or 18 carbons, has been chemically
attached (1). Partitioning of a compound occurs between the
hydrophobic stationary phase and a polar aqueous mobile phase.
Solvent systems usually consist of mixtures of methanol or
acetonitrile with water or buffer. Eluting strength increases
with decreases in polarity and, in general, compounds of similar
structure elute in order of decreasing polarity as can be seen for
a standard mixture (Figure 1a) of methyl, ethyl, propyl and butyl
parabens and the dye tartrazine (24). The four parabens have
relatively large alpha-values, where

$$\alpha = k'_2/k'_1$$

$$k' = (t - t_0)/t_0$$

where t is the retention time of a peak of interest and t_0 is
the retention time of an unretained substance. Such selectivity
for homologs is a characteristic of RPHPLC and has great potential
in the natural products area where mixtures of homologs are
frequently encountered. The highly charged tartrazine molecule,
which would stick irreversibly to a silica column, elutes at the
front. This demonstrates one of the beneficial aspects of RPHPLC,

namely, that very polar contaminants are not retained and therefore do not interfere with the chromatographic process. The fact
that aqueous solvents can be used is advantageous in processing
fermentation products where polar charged water-soluble materials
such as peptides are frequently encountered.
Other advantages of RPHPLC already well known in analytical
applications also carry over to preparative work. They include:
faster equilibration times on solvent changes, intrinsically
higher resolution and capacities than found on silica gel alone,
and the ability to handle a wide diversity of compounds through
the use of continuous or step gradients. As will be described
later, the use of such gradients allows for literally unlimited
injection volumes thus avoiding time-consuming lyophilization or
concentration steps.

Smaller scale preparative RPHPLC. Preparative chromatography can
of course be run on many scales. When confronted with a fermentation broth containing an unknown antibiotic, a first objective is
to isolate enough material to determine novelty. With today's
instrumentation [mass spectrometry (MS), infrared, ultraviolet
(UV), and nuclear magnetic resonance (NMR) spectroscopy], this can
often be done with 1-2 mg of material, readily prepared on
analytical or semi-preparative columns (4.6 or 10 X 250 mm) with
10 micron packing. However, for more detailed spectroscopic and
biological studies, larger amounts (several hundred mg) are
frequently required. For this scale we employ glass columns 2.5 X
50 cm (100 PSI limit) with larger particle supports. Because of
pressure limitations, flow rates are often limited to 15 ml/min.
Recently, several packings (Merck LiChroprep RP-18 25-40 micron,
Whatman Prep 40 ODS-3 C_{18} 37-60 micron or Baker Bonded Phase-
Octadecyl (C_{18}) 40 micron) have become available for such
columns. However, when gram quantities of products are required,
these glass columns are insufficient and larger instruments are
necessary as described below.

Materials and Methods

Analytical HPLC was run on a Beckman Model 345 Chromatograph
equipped with a Beckman 165 detector operated at 220 or 254 nm, as
indicated. Intermediate scale chromatography was run on Whatman
Prep 40 ODS-3 (37-60 micron) dry packed, or Merck LiChroprep RP-18
(25-40 micron) slurry packed (60% methanol-water) into glass
columns (2.54 X 50 cm, Altex-Beckman). Columns were equipped with
a pressure gauge (Ace Glass) and pressure release valves (Nupro)
set at 90 PSI. Elution was performed with an FMI metering pump
(maximum flow 15 ml/min at 100 PSI) equipped with an FMI pulse
dampener (Fluid Metering Company) or with a Beckman 112 HPLC pump
equipped with a preparative head (maximum flow 30 ml/min at 2000
PSI). Effluents were monitored with an ISCO UA-5 (Instrument
Specialties Company), or a Gow Mac Model 80-850 variable
wavelength detector (Gow Mac Instruments) at 210 or 254 nm.
Samples were introduced via a Rheodyne Teflon Valve (Rainin
Instruments), or by preadsorption of a solution of antibiotic onto

reversed phase packing or celite by concentration on a rotary
evaporator and packing onto a small Altex Beckman glass pre-column
(2.5 X 25 cm) equipped with a steel plunger. When solubility
permitted, larger volumes were injected through the pump in a
solvent containing lower amounts of organic modifier than required
for elution.

Larger scale chromatography was run on a J.Y. Chromatospac
Prep 100 (J.Y. Instruments) or a Waters Prep-500A equipped with a
Whatman Magnum 40 (4.8 X 50 cm) column dry packed with Whatman
Partisil Prep 40 ODS-3 (37-60 um) and mounted in place of the
radial compression chambers. Injections were carried out by
pumping on dilute solutions of samples in a solvent of lower
eluting power. Detection was by the Gow Mac model 80-850 U.V.
detector at 210 or 254 nm.

Burdick and Jackson acetonitrile (UV grade) was used for
analytical and glass column work. Baker acetonitrile (HPLC) was
used for larger scale work. Water for preparative work was
deionized and glass distilled.

Plate counts were determined by the formula

$$N = 5.54 \ (t/w_{.5})^2$$

where t is retention time and $w_{.5}$ is width at half height.
Values for plate counts are given as plates per column and not
normalized to plates per meter.

The polyoxin mixture was a water extract of the agricultural
product obtained from Kaken Kagaku Company. The other antibiotics
were originally unknown isolates from the SK&F fermentation screen.

Results

Low pressure glass column separations. Plate counts are normally
used in analytical columns as a measure of efficiency and
separation potential. We have found the determination of plate
counts to be desirable for the preparative columns in order to
monitor packing procedures and changes in column performance with
time. Using the parabens (Figure 1b) at analytical loading levels
on a glass column packed with 37-60 micron particles, 500 to 800
plates can be obtained. This compares with several thousand
plates for analytical columns with smaller particles (5 microns),
as seen in Figure 1a. These numbers should be used only for
comparative purposes, since plate counts tend to drop on increased
loading, (4-6,19) and the parabens tend to give higher plate
counts than complex antibiotics. Although seemingly low, 500
plates is sufficient to obtain adequate resolution when alpha-
values are 1.5 to 2.0.

Very tight separations (alpha values under 1.2) require
higher efficiency, and therefore can only be run with difficulty
on such large particle systems. When confronted with low alpha-
values, expensive preparative 10 micron columns (for example the
Whatman Magnum 20, 10,000 plates) run under non-overload conditions are required. Because such columns are small and operate
with high back pressures, they are limited both in loading

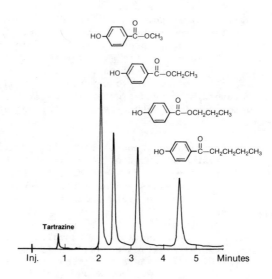

Figure 1a. Separations of parabens and tartrazine (0.1 μg each) on a preparative column (Beckman Ultrasphere ODS, 5 μm, 4.6 x 150 mm. Mobile phase: 60% methanol. Flow: 1 ml/min, 1700 psi. Detection: 254 nm (0.5 AUFS).

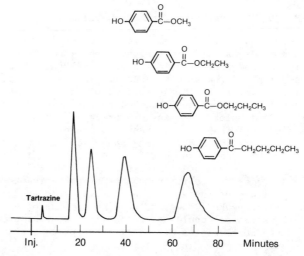

Figure 1b. Separations of parabens and tartrazine (0.5-1 mg each) on an analytical column (Whatman Partisil Prep 40 ODS-3, 37-60 μm, 2.5 x 50 cm). Mobile phase: 60% methanol. Flow: 20 ml/min. Detection: 254 nm (1 AUFS).

capacity and maximum allowable flow rate. Furthermore, these high
resolution, small particle columns suffer very severe plate count
drops on excessive loading. Thus, the plate count of 10,000 does
not carry over to large scale preparative work. In order to get
high throughput, the larger particle, lower resolution systems are
necessary. Thus, higher alpha-values need to be obtained either
by varying solvent, pH or packing chemistry, or by adjusting
fermentation parameters to remove offending materials. If none of
these can be done, scale-up with reasonable throughput will be
very difficult.

Examples. Using the glass systems on partially purified prepara-
tions, we have separated mixtures of polyoxins L, B, A and K,
[(25), Figure 2] and gilvocarcins V and M [(26), Figure 3]. The
polyoxins are polar nucleosides and originally required extensive
chromatography on cellulose to yield products for chemical
characterization and could not be chromatographed on silica (25).
Using paired ion chromatography with hepta-fluorobutyric acid
(HFBA), the four major components of the complex which differ in
amino acid content were readily separated in 100 mg quantities
(Figure 4). Structures of these antibiotics were assigned by fast
atom bombardment mass spectrometry and NMR (27).
    In a second example of preparative RPHPLC, an ethyl acetate
extract of an unknown antibiotic which had been chromatographed on
silica was further fractionated into homologs (Figure 5).
Analysis by UV, MS and NMR indicated that these materials were
gilvocarcins M and V, differing only in the presence of a methyl
or vinyl side chain (26).

Separation of a Glycopeptide Antibiotic Mixture

The isolation and separation of three glycopeptide antibiotics
(28) from a crude XAD-7 fermentation broth extract further
exemplifies the advantages of preparative RPHPLC. Figure 6a gives
an analytical chromatogram of a crude isolate of a mixture of
antibiotics of interest. The antibiotic was not solvent extractable and could not be chromatographed on silica. In exploratory
work with this complex, samples were processed through several
purification steps (XAD, ion exchange, Sephadex, LH-20) and
finally on preparative RPHPLC using analytical and glass preparative columns. It was shown that the peaks labeled I, II and III
were the antibiotic components of interest. The multistep
sequence was necessary to remove the polar contaminants which
elute at the front of the analytical chromatogram and which tail
into the desired peaks. However, gram quantities of each antibiotic were needed and could not be obtained by that process
because of the complexities of the clean-up procedure and the
limited capacity of the glass columns.

Design of the separation. It is well known that gradients can
improve apparent resolution in chromatography (1). The use of a
gradient in a separation such as that shown in Figure 6a would be
ideal in that adequate clearance of the polar contaminants could

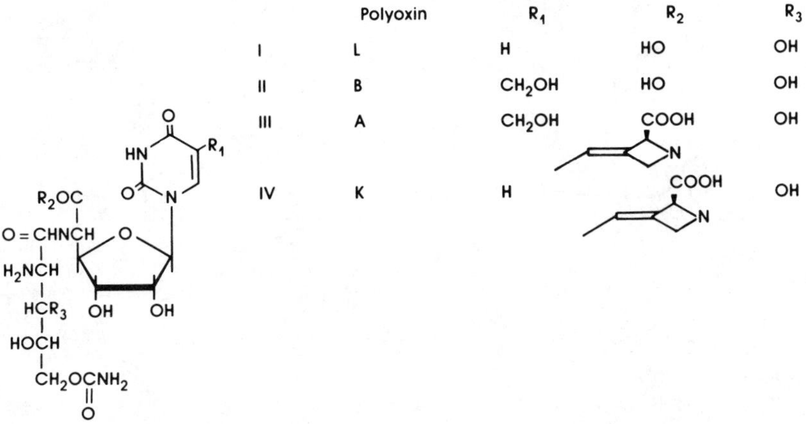

	Polyoxin	R_1	R_2	R_3
I	L	H	HO	OH
II	B	CH_2OH	HO	OH
III	A	CH_2OH	COOH (structure)	OH
IV	K	H	COOH (structure)	OH

Figure 2. Structures of polyoxins L, B, A, and K.

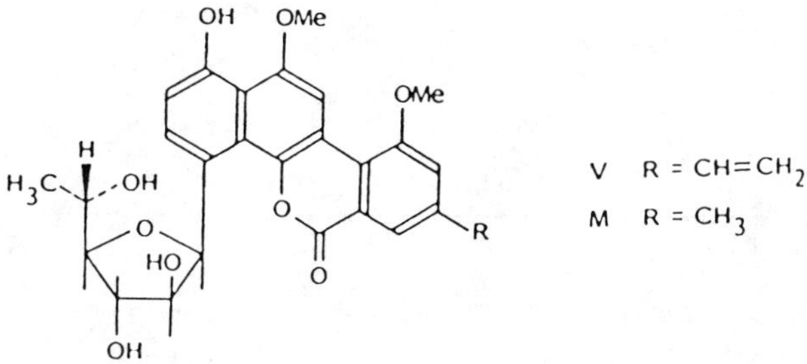

V   R = CH=CH₂

M   R = CH₃

Figure 3. Structures of gilvocarcins V and M.

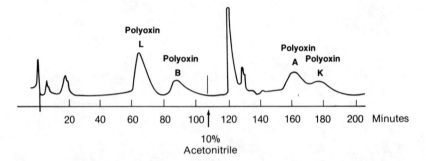

Figure 4. Separation of a polyoxin mixture. Sample: polyoxin
extract (100 mg). Column: LiChroprep RP-18, 25-40 μm, 2.5 x
50 cm. Mobile phase: 0.015 M HFBA. Flow: 14 ml/min, 80 psi.
Detection: 254 nm.

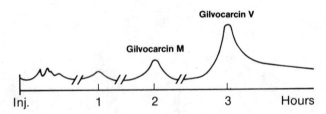

Figure 5. Separation of a gilvocarcin mixture. Sample: gilvo-
carcin (40 mg). Column: LiChroprep RP-18, 25-40 μm, 2.5 x 50 cm.
Mobile phase: 35% acetonitrile. Flow: 8 ml/min, 80 psi.
Detection: 254 nm.

take place before elution of Peak I. Figure 6b shows an
analytical chromatogram of the same crude material eluted with a
gradient of acetonitrile. In this case the front material is
completely resolved from the desired peaks. When 2 g of crude
complex was chromatographed on a glass column using a step
gradient a substantial clean-up was observed, but resolution and
throughput were still limited by column size and flow rate
limitations (See Figure 7). This separation took over 6 hours and
yielded minimal amounts of products. Scaling up to larger
instruments was necessary.

Scale-up instrumentaiton. Two commercially available instruments
were used for the scale-up work, a J.Y. Chromatospac 100 and a
Waters Prep-500A equipped with a Whatman Magnum 40 column, as
described in Materials and Methods. The JY unit uses an axial
compression system to obtain tight packing. In our hands, both
systems displayed equivalent resolution. The Waters unit with the
Whatman column had slightly lower capacity because its column was
smaller, but solvent changes and sample injections were easier to
carry out with its reciprocating pump. In both cases, a Gow Mac
variable wavelength UV detector was used at 210 or 254 nm. This
detector is very desirable as it shows a linear response to very
concentrated solutions of compounds at their optimal absorbances.
Detection of the desired material is enhanced since this detector
can be used at the same wavelength as that used in corresponding
analytical work. In our hands, the use of a normal variable
wavelength detector run at a wavelength somewhat removed from the
maximum, as often recommended, has frequently given misleading
results due to the numerous UV-absorbing contaminants in fermentation products. The cell of the Gow Mac detector has a 0.1 mm
pathlength making it 100 times less sensitive than an analytical
UV detector with a 10 mm pathlength. The cell design can also
handle 500 ml/min, the maximum flow rate of both large scale
chromatographs.
    Both large scale systems displayed lower resolution (200-300
plates) than the glass columns when evaluated with the parabens,
but they still had sufficient resolving power to separate the
paraben mixture (Figure 8). In theory, efficiency is still
sufficient for resolving the three major components in Figure 4
(alpha's of 1.6 and 1.9). Using alpha = 1.6, N = 300, and k'= 10
(typical values from Figure 4 and conditions used for preparative
work) and the well-known ($\underline{1}$) resolution equation:

$$R_s = 1/4 \ (\alpha-1)\sqrt{N} \ k'/(k'+1)$$

where R_s (resolution) is the ratio of the distance between two
peaks divided by their average band width, a calculated resolution
of 2.3 is achieved. Such a value implies baseline resolution ($\underline{1}$)
and is well in excess of that needed to resolve peaks I and II
(Figure 6a). (The minor peaks have relative alpha's of only 1.25
implying an R_s = 0.96, which is still sufficient if center cuts
are taken.) These calculations use plate counts determined for
ideal substances under non-overload conditions. For conditions

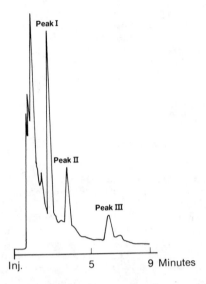

Figure 6a. Analytical separations of a crude glycopeptide complex
(isocratic). Sample: crude antibiotic extract. Column: Beckman
Ultrasphere ODS, 5 μm, 4.6 x 150 mm. Mobile phase: 35% acetonitrile in 0.1 M KH_2PO_4pH 3.2. Flow: 1.5 ml/min. Detection:
220 nm.

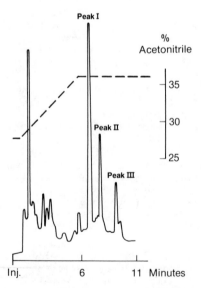

Figure 6b. Analytical separations of a crude glycopeptide complex
(gradient). Sample: crude antibiotic extract. Column: Beckman
Ultrasphere ODS, 5 μm, 4.6 x 150 mm. Mobile phase: 27 to 37%
acetonitrile in 0.1 M KH_2PO_4pH 3.2. Flow: 1.5 ml/min. Detection:
220 nm.

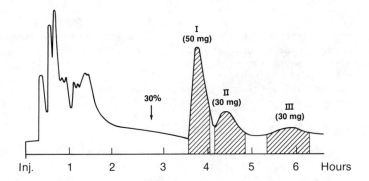

Figure 7.   Small-scale preparative separation of glycopeptide
complex.   Sample: crude antibiotic extract (2 g).   Column:
Merck LiChroprep RP-18, 25-40 μm, 2.5 x 50 cm.   Mobile phase: 25
to 30% acetonitrile in 0.1 M KH_2PO_4pH 3.2.   Flow: 14 ml/min,
90 psi.   Detection: 210 nm (Gow Mac).

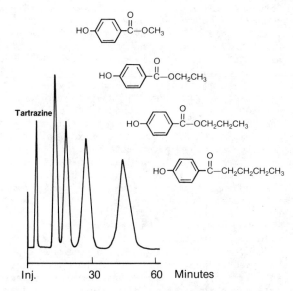

Figure 8.   Separation of parabens on Magnum 40 column (4.8 x 50
cm).   Sample: parabens and tartrazine (60-80 mg each).   Column:
Whatman Partisil Prep 40 ODS-3 (37-60 μm).   Mobile phase: 60%
methanol.   Flow: 100 ml/min.   Detection: 254 nm.

run at high loading ($>$10 mg/g) lower plate counts are observed but
often other beneficial effects such as those observed in displacement chromatography become evident (18).

An example of a preparative separation of a purified fermentation extract in order to obtain suitable elution conditions using
the Magnum column, is shown in Figure 9a. The sample (2 g), was
injected through the pump in a solvent system lower in acetonitrile content than needed to elute Peak I. Elution with a step
gradient yielded (after desalting) each of the pure components
with better than 90% recovery and purity (see Figure 9b). Loading
in this case was approximately 1 mg of Component I per gram of
adsorbent.

Factors Potentially Limiting Scale-up

With suitable solvent systems worked out, the question arose as to
what extent this process could be scaled up to produce gram quantities of each component. Three potential limitations existed:
1) Supplies of purified intermediate, 2) loading capacity of the
column, 3) size of column and cost of packing.

Bypassing complex isolation scheme. The multistep procedure used
to prepare the starting material for this chromatographic separation was found to be the primary bottleneck. Therefore, in the
next scale-up experiment, 25 g of a crude XAD isolate (containing
2.2 g of Component I) in 4 liters of 17% acetonitrile in buffer
was pumped onto the column at 250 ml/min. Sequential step elu-
tions (20, 22, 24 and 26% acetonitrile at 250 ml/min) resulted in
nearly baseline resolution of the three components (Figures 10a
and 10b) again with high recovery and purity ($>$85%), in less than
3 hours. Since separate experiments had indicated that acceptable
resolution could be obtained at higher flow rates, all scale-up
work was performed at 250 ml/min. Thus, the preparative reversed
phase column not only yielded chromatographically pure products
but could achieve the preliminary purification with high recovery
in hours, a process which otherwise took several days to do.

When 50 g (containing approximately 3 g of each component) was
injected in 6 liters of buffer alone, no breakthrough was observed,
and on gradient elution, the chromatogram shown in Figure 11a was
obtained. Although resolution between the three components had
deteriorated, the complex had been efficiently separated from the
contaminants in less than 2 hours. This first chromatography step
gives a "window" cut which is relatively free of contaminants of
higher and lower polarity and thereby more amenable to rechromatography. Indeed, on rechromatography, a superior separation of the
complex into its components occurred. The poorly resolved mixture
of Components I, II and III from Figure 11a was pooled, diluted
with buffer and rechromatographed to yield, after desalting, 3 g
each of the pure components (Figure 11b). Using the above two-
step procedure (i.e. several separations on 50 g or more of crude
extract followed by rechromatography of the pooled purified
complex) over 10 grams of the major component was obtained, with
the same high recovery and purity as shown in Figure 9b.

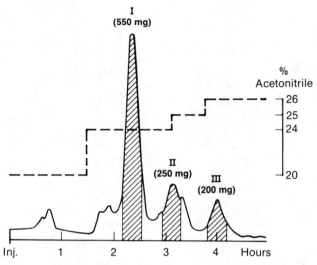

Figure 9a. Small-scale preparative separation of glycopeptide complex on Magnum 40 column (4.5 x 50 cm). Sample: purified antibiotic (2 g). Column: Whatman Partisil Prep 40 ODS-3 (37-60 μm). Mobile phase: 20 to 26% acetonitrile in 0.1 M KH_2PO_4pH 6.0. Flow: 100 ml/min. Detection: 210 nm.

Antibiotic Component	Initial Content	Isolated Weight	HPLC Purity
I	600 mᵧ	550 mg	>95%
II	350 mg	250 mg	>90%
III	250 mg	200 mg	>95%

Inj.

Figure 9b. Recovery data for small-scale separation of glycopeptide complex on Magnum 40 column. Sample: purified component I. Column: Beckman Ultrasphere ODS, 5 μm, 4.6 x 150 mm). Mobile phase: 27 to 37% acetonitrile in 0.1 M KH_2PO_4pH 3.2. Flow: 1.5 ml/min. Detection: 220 nm.

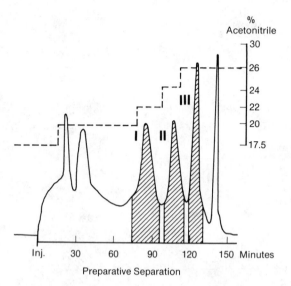

Figure 10a. Scale-up preparative separation of glycopeptide complex (25 g run). Sample: crude antibiotic complex (25 g). Column: Whatman Partisil Prep 40 ODS-3 (37-60 μm), Whatman Magnum 40 (4.8 x 50 cm). Mobile phase: 17.5 to 26% acetonitrile in 0.1 M KH_2PO_4 pH 6.0. Flow: 250 ml/min. Detection: 210 nm.

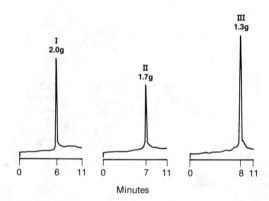

Figure 10b. Analysis of fractions for scale-up separation of glycopeptide complex (25 g run). Sample: pools from 25 g run. Column: Beckman Ultrasphere ODS (5 μm), 4.6 x 150 mm. Mobile phase: 27 to 37% acetonitrile in 0.1 M KH_2PO_4 pH 3.2. Flow: 1.5 ml/min. Detection: 220 nm.

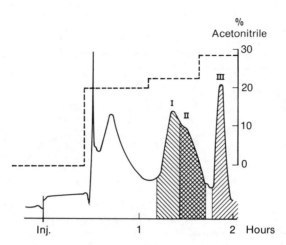

Figure 11a. Scale-up preparative separation of glycopeptide complex (50 g run). Sample: crude antibiotic complex (50 g) in 6 L
buffer. Column: Whatman Partisil Prep 40 ODS-3 (37-60 μm), Whatman Magnum 40 (4.8 x 50 cm). Mobile phase: 0 to 28% acetonitrile
in 0.1 M KH$_2$PO$_4$pH 6.0. Flow: 250 ml/min. Detection: 210 nm.

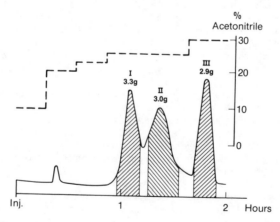

Figure 11b. Rechromatography of pooled fractions for scale-up
separation of glycopeptide complex (50 g run). Column: Whatman Partisil Prep 40 ODS-3 (37-60 μm), Whatman Magnum 40 (4.8 x
50 cm). Mobile phase: 10 to 28% acetonitrile in 0.1 M KH$_2$PO$_4$pH
6.0. Flow: 250 ml/min. Detection: 210 nm.

The use of the column as a concentrator is notable. Pools containing relatively large volumes of fractions (10-20 liters or larger) can be mixed with equal volumes of buffer to dilute out the acetonitrile and pumped onto the column with complete retention. Normal elution can give the desired chromatographic separation or, using a very strong solvent, the entire product can be concentrated into two to three liters, a volume suitable for desalting, lyophilization, etc. Such a step takes less time than lyophilization or evaporation of the entire volume.

Loading capacity. The second factor which limits throughput, indicated earlier, is the loading capacity of the column. Table I shows the results of a loading study with homogeneous Compound I which indicates that "overload" (as defined (1) by a 10% loss in k') occurs at levels of 1-2 mg of pure material per g adsorbent. The k' values in Table I are apparent k's since the chromtographic procedure uses a step gradient in order to inject large volumes. Actual k's are somewhat lower. Also, loss of resolution (defined by N) occurs as loading increases beyond that level. On recalculation of resolution observed at the highest level tested (13 mg/g) using N = 53, alpha = 1.6 and k'= 16, a minimally acceptable value of 1.0 is obtained. In theory, this probably represents the maximum loading for separation of Component I from Component II. At present, quantities are not available to confirm this. Although substantially higher loadings are possible by using displacement chromatography, the overall throughput would still depend on the capacity of the column to purify the larger amounts of crude extracts needed to prepare the precursors for the displacement chromatography step. The maximum loading of crude extract is yet to be determined because of supply considerations.

Cost limitations. The third limitation to scale up is the size of column used and cost of packing. In carrying out large scale work on crude extracts, the cost of packing is dependent on column life. We have found that the packing turns dark in color and loses resolution after several 50 g runs. However, it can be completely regenerated by extrusion, soaking in DMSO and repacking. Because the packing can be regenerated and reused, its modest cost ($700/kg) is not a serious obstacle to its use on crude material. Should further scale-up be required, larger columns with 4 and 20 times the capacity as the Prep-40 column are available from Whatman, as are the industrial units from Whatman and Waters.

Conclusions

In summary, we have demonstrated the potential utility of preparative reversed phase HPLC as a tool to purify fermentation products in multigram scale on laboratory instruments. Secondly, due to its intrinsic selectivity and compatibility with the use of high flow rates, RPHLC is a useful tool for rapidly purifying crude fermentation extracts to a level of purity equivalent to that obtained by multistep time consuming procedures. We

TABLE I. Loading Study - Component I[a]

Injection					
Weight (g)	Volume[b] (ml)	Loading (mg/g)	k' [c]	N	R_s [d]
0.2	80	0.3	19.2	147	1.7
0.2	2000	0.3	19.3	142	1.7
0.6	220	1.0	17.8	122	1.6
1.2	440	2.0	17.7	121	1.6
2.4	880	4.0	17.2	95	1.4
5.0	1900	8.3	16.2	61	1.1
7.7	1000	12.8	17.4	51	1.0

a) Eluant: 24% CH_3CN in 0.1 M phosphate buffer pH 6.0 at 200 ml/min. Whatman Prep 40 ODS-3 Packing (37-60 micron) in Magnum 40 Column 50 X 4.8 cm; UV detection at 210 um.

b) Sample dissolved in buffer, no acetonitrile. Column was pre-equilibrated with same buffer.

c) Apparent k', not corrected for equilibration time.

d) Calculated using alpha = 1.6.

recommend a two-step procedure to get maximum throughput of
material: 1) Several runs, loaded as heavily as possible with
crude extracts to get a "window" containing the desired
material(s), 2) rechromatography of pooled cuts to produce the
pure product.

Finally, the overall loss in plate count in going from a 5
micron analytical column run at low loading with ideal substances
to a larger particle preparative column (37-60 microns) run at
high loading with complex substrates is indeed striking.
Nevertheless, for preparative purposes such resolution is still
sufficient to separate compounds cleanly with high recovery
provided that alpha-values are large enough (>1.5). It would
appear that excessive concern with plate counts in such systems is
unwarranted unless very close separations are being run.

Acknowledgment

We gratefully acknowledge the excellent technical assistance of
Mr. George Udowenko and the support of other members of the SK&F
Department of Natural Products Pharmacology for supplies of
fermentation products. We thank Dr. James Chan for data on the
structure of the purified polyoxins.

Literature Cited

1. Snyder, L.; Kirkland, J. "Introduction to Modern Liquid
    Chromatography", Second Edition, John Wiley, New York, 1979.
2. Kabra, P.; Maron, L. "Liquid Chromatography in Clinical
    Analysis", Humana Press, Clifton, NJ, 1981.
3. Poppe, H.; Kraak, J. J. Chromatog. , 1983, 295, 395.
4. De Jong, A.; Poppe, H.; Kraak, J. J. Chromatog. 1981, 209,
    432.
5. De Jong, A.; Poppe, H.; Kraak, J. J. Chromatog. 1978, 148,
    127.
6. De Jong, A.; Kraak, J.; Poppe, H.; Nooitgedacht, F. J.
    Chromatog. 1980, 193, 181.
7. Hupe, K.; Lauer, H. J. Chromatog. 1981, 203, 41.
8. Coq, B.; Cretier, G.; Rocca, J. Anal. Chem. 1982, 54, 2271.
9. Sugnaux, F.; Djerassi, C. J. Chromatog. 1982, 248, 373.
10. Rabel, F. Am. Lab. 1980, 12, 126.
11. Hupe, K.; Lauer, H.; Zech, K. Chromatographia 1980, 13, 413.
12. Verzele, M.; Dewaele, C.; Van Dijck, J.; Van Haver, D. J.
    Chromatog. 1982, 249, 231.
13. Kagan, M.; Kraevskaya, M.; Vasilieva, V.; Zinkevich, E. J.
    Chromatog. 1981, 219, 183.
14. Fiedler, H. J. Chromatog. 1981, 209, 103.
15. Horvath, C.; Nahum, A.; Frenz, J. J. Chromatog. 1981, 218,
    365.
16. Kalasz, H.; Horvath, C. J. Chromatog. 1981, 215, 295.
17. Kalasz, H.; Horvath, C. J. Chromatog. 1982, 239, 423.
18. Horvath, C.; Frenz, J.; Rassi, Z. J. Chromatog. 1983, 255,
    273.

19. Snyder, L. and Kirkland, J., <u>ob. cit.</u>, Chapter 15,
    "Preparative Liquid Chromatography".
20. Nettleton, D. <u>J. Liq. Chromatog.</u> 1981, 4 (Supp. 1), 141.
21. Haywood, P.; Munro, G. In "Developments in Chromatog-
    raphy - 2"; C.F.H. Knapman, Ed.; Applied Science Publishers,
    London 1981, p.33.
22. Eisenbeiss, F.; Henke, H. <u>J. of High Res. Chromatog. and
    Chromatog. Comm.</u> 1980, 2, 733.
23. Gasparrini, F.; Cacchi, S.; Cagliotti, L.; Misiti, D.;
    Giovannoli, M. <u>J. Chromatog.</u> 1980, 194, 239.
24. Merck Index, 10th Edition, M. Windholz, Ed., Rayway, NJ. 1983,
    p1303.
25. Isono, K.; Asahi, K.; Suzuki, S. <u>J. Am. Chem. Soc.</u> 1969, 91,
    7490.
26. a) Nakano, H. et al. <u>J. Antibiotics</u> 1981, 34 266;
    b) Takahashi, K. et al. <u>Ibid</u> 1981, 34, 271.
27. Chan, J.; Yeung, E.; Roberts, G.; Sitrin, R. unpublished data.
28. Sitrin, R.; Chan, G.; DeBrosse, C.; Dingerdissen, J.; Hoover,
    J.; Jeffs, P.; Roberts, G.; Rottschaeffer, S.; Valenta, J.;
    Snader, K.; <u>24th Interscience Conference on Antimicrobial
    Agents and Chemotherapy</u> 1984, Abs. No. 1137.

RECEIVED September 7, 1984

Process Scale Chromatography
The New Frontier in High Performance Liquid Chromatography

A. H. HECKENDORF[1], E. ASHARE, and C. RAUSCH

Waters Associates, Inc., Milford, MA 01757

In a process scale chromatograph, the concern for
continuous utilization of an expensive investment in
capital equipment plays a key role in the trade-offs of
operation. The scale-up from the laboratory to the
industrial scale equipment requires a system engineered
to solve a problem greater than the task of purifying
larger amounts of compound. The problem is one of
operating a system at minimum cost and obtaining high
yields of compound at high purity in as short a period
of time as possible, and at as high a concentration as
possible to minimize cost and solvent removal problems
in the recovery process. The use of multiple column
segments allows the flexibility of tailoring the column
length to the difficulty of varieties of separations and
through column sequence and dynamic column length
control, many different modes of operation can be
performed.

High performance liquid chromatography has developed from the
analytical scale to the process scale. The evolution of column
technology was enhanced by a technique of radial compression, first
developed in 1975. This technology was the result of the
realization that rather than go to smaller and smaller particle
technology to gain resolution in order to enhance the ability of a
packed bed structure to resolve compounds, the spaces between
particles could be decreased, thus decreasing the amount of
dilution that would occur from a given mass transfer rate. This
significant step in the evolution of analytical HPLC technology
resulted in the ability to change direction away from open column
chromatography to the utilization of high-speed, high-resolution,
column technology.

[1]Current address: NEST Group, Southboro, MA 01772.

Introduction to the Technology of Chromatography

If one compares the efficiency or resolving power of smaller particle materials in a packed bed structure to the load requirements of preparative chromatography (Figure 1), one can see that, in fact, resolution decreases with load. Hence, the drive toward smaller particles is a goal for the analytical chromatographer, but it is not a mutually shared goal for the preparative chromatographer. Because prep is the relative enrichment of one compound vs. another, it is the ability to isolate compounds from a packed bed structure in high yield and high mass in a short period of time.

This Radial Compression technology has been packaged in a variety of forms, the smallest of which is the SEP-PAK cartridge for sample preparation for the analytical chromatographer. A SEP-PAK cartridge illustrates how column technology can function much like a liquid-liquid extractor, while utilizing an immobilized bed. The SEP-PAK C-18 cartridge is filled with silica gel which is covalently bonded with an octadecyl silane ether. This C-18 (octadecyl) coating functions like the non-polar solvent in a liquid/liquid extractor, and if one passes a polar solvent across this column, a partition phenomenon occurs. If one needs to separate the components of a mixture, one can select solvent conditions which will optimize the partitioning ability of the non-polar immobilized phase. If one uses water, the furthest in polarity away from the non-polar C-18 group, neutral polar and non-polar sample components are partitioned into the surface coating. If the amount of miscible non-polar solvent in the mobile phase is selectively increased, one can selectively elute or partition away from that bonded phase the various components. If one increases even further the non-polar component of the mobile phase, the additional components begin to elute. Thus, rather quickly we can fractionate a complex mixture into several components, or fractionate a simple mixture into its individual components.

Column Technology Design

The technology offered to the process industry is a combination of column technology, column chemistry and system architecture optimized around a particular task. The key element to most chromatographic systems is based on column design, and the parameters for column performance are optimized. If one were concerned about the effectiveness of the packed bed structure, and in keeping it in place for long periods of time, and one wanted to minimize the use of pressure because the trend is towards higher pressures to get better resolution, one needs an alternate approach to scaling up column capacity.

One can get that better resolution for a particular particle size without the increased pressure through radial compression technology. The need to improve resolution as in the analytical world is there, but in the preparative world there is also the need to control the bed structure over long periods of time. The use of massive overload and massive overworking of the column to serve production purposes, demands a more rigid control over the

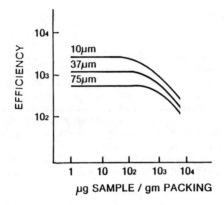

Figure 1. Load vs. particle size.

potential channeling and voiding of that bed structure through
solvent changes, solvent stripping, and general abuse that occurs
in the process environment to control costs. If one can keep
pressures low, one also can increase the speed with which one can
flow through that column and, thus, the separation time is going to
be at a minimum. One wants to get the best efficiency out of that
bed structure that one can, regardless of that fact that one will
lose a good portion of that efficiency through loadability. And,
one is going to want to have the best capacity for separating
compounds that one can with a fixed bed length.

$$R = 1/4 \ \sqrt{N} \ \ \frac{\alpha-1}{\alpha} \ \ \frac{k'}{k'+1}$$

The resolution equation is composed of three parts: the efficiency
parameter, N, the chemistry or selectivity parameter, alpha, and
the capacity or loadability parameter, k'. In the resolution
equation for the preparative environment, this equation is already
fixed by the chemistry parameters. For example, the capacity
factor is set for the optimum load from the analytical data; the
chemistry or selectivity parameter is set by the analytical work-up
where one has tested all the possible combinations of chemistry
required to get the separation; and the efficiency parameter is
fixed by the amount of load that is going to be put on that bed
structure.

The efficiency, N, is a function of the length and the height
equivalent to a theoretical plate.

$$N = L/H$$

The height equivalent to a theoretical plate is composed of three
components of the equation:

$$H = A + B/\mu + C\mu$$

The A term (Eddy diffusion) is the term minimized by radial
compression technology; the B term is positively influenced by
linear velocity increases. If one wants to go to higher speeds,
the longitudinal diffusion (the B term) gets smaller. This is good
because one is going to want to use the system at as high a rate as
possible to get the maximum throughput out of this investment.
However, the C term is negatively affected by this increase in
linear velocity, μ, So one needs some fluid velocity control to be
able to optimally get the separation over the widest possible
operating range. Thus, the goals are to drive the H term down (the
height equivalent to a theoretical plate down), get the optimum use
of pressure, and to maintain H over a wide dynamic flow range, so
that one can do a number of different separations on the same
system at different linear velocities.

If one considers the effects of packing material particle size on linear velocity, it has been determined that if one decreases particle size by a factor of two, the pressure increases as its square. Thus, the productivity increase is potentially jeopardized for an esthetic increase in resolution. One needs to test the actual resolution achieved in chromatograms of "unresolved peaks" by taking fractions through the region of interest and analyzing their composition.

Analogously, the choice of pore size on productivity must be evaluated too. The recent interest in large pore (greater than 300 angstrom) packing materials for chromatography of macromolecules such as peptides and proteins was initiated to allow higher recoveries from each chromatographic run, as well as to eliminate "memory" effects from entrapped molecules on subsequent runs. However, the larger the pore size, the smaller the surface area becomes. This results in lower capacity factors, k', as well as lower total capacity from a chromatographic system. Thus, testing of relative loadability and capacity from the packing materials is necessary to optimize the productivities of any system.

A significant consideration in the production environment is the concentration of the product in the eluent due to the costs associated with sample recovery systems and compound stability during recovery. Concentration is a function of the length and diameter of the column. The wider the column and the longer it is, the more dilute the product stream is going to be. Thus, the more energy and time required to recover the components of interest, and the higher the potential for product yield losses due to degrada-tion of purified product. Thus, consideration of techniques to prevent product dilution in a chromatographic system are important.

The amount of time involved with the separation is going to be a function of the length of the column, the linear velocity, and the capacity term we discussed above. If one wants to get the most flexible column system for a variety of separation problems, what one is going to want to do is optimize each of these terms (fluid velocity, concentration and separation time). What one is forced to consider is the ability to have a dynamic column length control built into the system architecture, which will minimize the amount of dilution, shorten the amount of time to get components out, keep the pressure at its lowest possible level for maximum operating capability, and not sacrifice the ability to separate compounds. If one incorporates radial compression technology into the chromatographic system, it offers one the ability to segment the column into short segments without sacrificing fluid distribution or increasing product dilution through this segmentation of the bed; while, at the same time, it offers the ability to get the maximum efficiency of our sample feed rate with respect to the column length.

In Figure 2, one can see that the ratio of the column length available over the minimum length necessary to get a separation, plotted against the efficiency of the sample feed rate, goes through an optimum. And, depending on load, this optimum can shift, so that the need for having a variable length column within a single system is apparent, especially when one compares that same L over L min. ratio against the efficiency of solvent utilization.

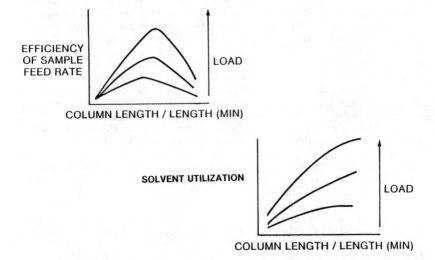

Figure 2.   Effects of dynamic column length control.

If one can fix the column length at an optimum for the sample feed rate, then one can fix the amount of solvent required to get that separation.

Fluid Velocity Control

One must look at the factors affecting sample distribution to see how one can control this fluid velocity parameter, to allow segmentation of the column. Scaling parameters for sample distribution involve fluid distribution, control over temperature and, as has been discussed, a pressure consideration as well as a mixing volume concern.

If one does not control mixing volumes adequately, there will be an automatic increase in the volume of the product. If one utilizes a large volume taper at the end of the column to control fluid velocity, as has been done in the laboratory column technologies, one can get a smooth addition of sample onto the column at low linear velocities. However, in a production environment where one is going to try to optimally pump that bed structure, one can see from the Van Deemeter plot comparison to the fluid velocity profiles within this schematic of a column (Figure 3) that one will be operating at different linear velocities within that distribution orifice. This adds volume to the product, and consequently, there's a loss of resolving power within the system.

Considering the effect of momentum on the volume of the peak through this schematic here, one had a point source of sample and solvent at high linear velocities, the center of the peak would be driven downwards within the column because of momentum. If one uses a distribution plate to interrupt that momentum and to adequately distribute the sample across the bed structure, less of a momentum parameter will be introduced to that separation. This latter situation can be achieved most effectively if one has eliminated the wall effect through radial compression technology, and has obtained a uniform density bed through the same technology. Schematically, this fluid distribution system has been achieved on our technology through the use of a set of thin, low volume disks, supported by the force of the axial bridging of the packing under radial compression, so that the sample is put on in a uniform manner, regardless of flow rate through the bed.

Schematically, one can get a variety of band profiles from column design and bed structure parameters. On the top left of Figure 4 is one profile from a high linear velocity flow of the column. The center profile shows the effect of wall effect and point source distribution of sample and solvent at high linear velocities. If one eliminates the wall effect by utilizing very wide diameter columns and smaller particle packing, but still has a point source distribution of sample and solvent, one will get band distortion in the center of the band at higher fluid velocities. The net result in a process environment would be a broad dilute product peak. If one can combine wall effect, bed density and fluid distribution control, one can narrow this band and increase the concentration of product, as illustrated on the bottom of Figure 4.

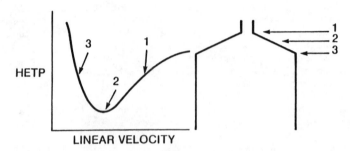

Figure 3. Column efficiency vs. column geometry.

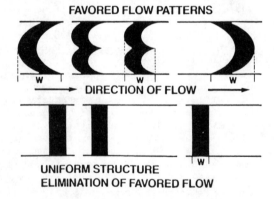

Figure 4. Band spreading.

System Architecture

Having designed an optimum column for separation, we can now
incorporate this technology into the optimum system architecture.
This system architecture should incorporate the segmented column
design that is necessary for a maximally flexible unit separation
process in a production environment. Segmented column design
allows column sequence control within the system. Thus, one can
get a more continuous recovery of product or a higher operating
efficiency out of the system. And, because one is segmenting that
column, one can use more of the column bed for an operation at any
single time than would be able to be done with a single large
column.

Schematically (Figure 5), one can compare a separation of two
components—one on a long column and one on a short one—and one
can see that, with a segmented design one can pull out into the
product stream a material that is partially purified or material
that is fully purified, and the partially purified material is then
sequenced onto another column segment for further purification.
This minimizes the amount of dead time between sample stream
additions and also allows one to view what's going on within the
column at a higher frequency rate than one is able to do on a
larger column.

Through parallel column segment operation (Figure 6), one can
continually load product onto the columns at shorter cycles, thus
getting a higher product stream output.

If one is going to control column sequence, this allows one
to utilize complex solvent sequences as well (Figure 7). For
example, for certain samples one will have to put on a sample;
after that separation is complete, one will have to flush that bed
structure and then one will have to re-equilibrate that bed
structure before the next sample can be added. If one uses a
segmented column system, one can operate each column independently
in terms of the operation sequence, and optimally get the
downstream side of the system functioning at its maximum because,
at each step, there will be a product coming out of the bed.

In a three column system, on Column One the column
operational sequence will be sample, flush, equilibrate, and then
on Column Two one will have to equilibrate, sample, flush; and to
set up Column Three in sequence, one is going to equilibrate
through the first two steps and then load sample and then flush.
Note that the product is continuously coming out of the system
after each sequence. This is different than if one had a single
column and one had to wait for longer periods of time between
events. It allows optimum utilization of downstream recovery
equipment with less batch operation.

Dynamic column length control and column sequence control
also offer the ability to fractionate complex mixtures by the use
of various column segments to perform the separation tasks
differently. For example: In the configuration in Figure 8, one
can flush the early eluters from a complex mixture, and then shunt
the partially purified product onto the separation segments of the
system. And simultaneously, flush the long eluters off that first
column segment as well. This gives one some flexibility in
operation by allowing the system to handle both complex and simple
separation processes on the same equipment.

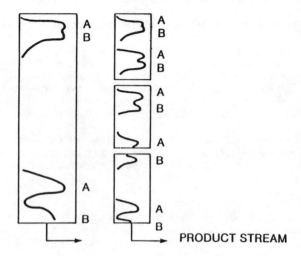

Figure 5.   Column segmentation for optimum column utilization.

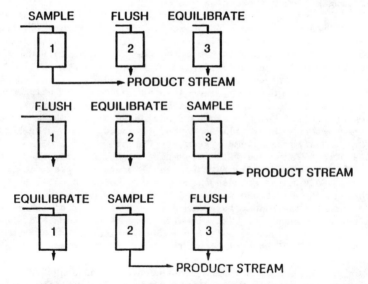

Figure 6.   Column sequence control (for complex solvent sequences.)

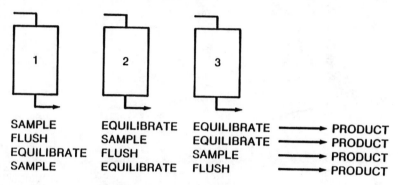

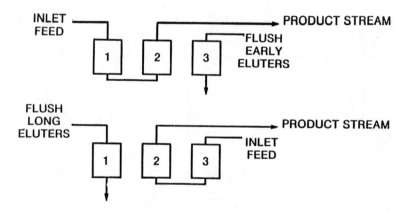

Figure 7.   Column sequence control (for complex solvent sequences.)

Figure 8.   Dynamic column length control and column sequence control (for complex mixture simplification.)

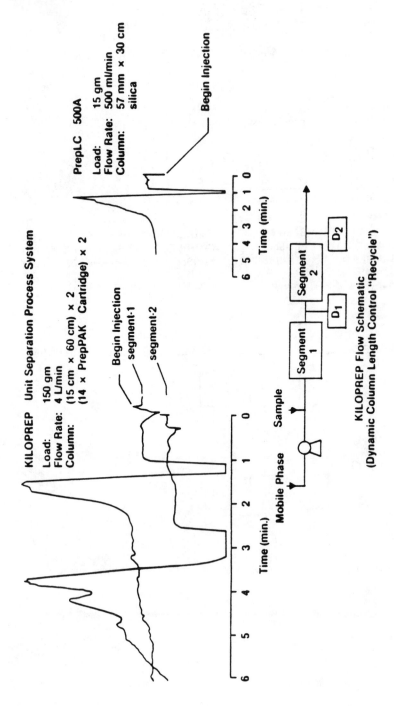

Figure 9. Comparison of column performance at equivalent loading.

This technology has been built into the Waters pilot plant
unit called the KILOPREP   Process Chromatography System. It's
the first of a series of increasingly larger chromatographs
incorporating radial compression technology into the system. The
performance of this larger system can be predicted from the data
generated in an equivalent laboratory device as in Figure 9. This
is important to process development programs because it minimizes
the amount of product and solvent needed to work out a development
program.

The technology of high performance liquid chromatography has
been successfully extended from the analytical scale to the process
scale. The ability to control the various operation parameters to
scale up directly from the laboratory to the pilot plant and beyond
to the production environment has been developed. This technology
can be combined with other separations technologies, such as
membrane separations, to provide particle-free solvents, ultrapure
products, and concentrated product streams. This will give the
opportunity to deal with future separations problems of the
chemical process industry.

RECEIVED October 4, 1984

Immunosorbent Chromatography for Recovery of Protein Products

JOHN P. HAMMAN and GARY J. CALTON

Purification Engineering, Inc., Columbia, MD 21046

The recovery of protein products from fermentation
processes by immunosorbent chromatography can be
economically attractive. The amount of immunosorbent
required is a major cost factor. The proper selection
of antibody, matrix, immobilization method and elution
conditions can allow high throughput for a given volume
of immunosorbent. The throughput depends mainly on the
flow rates through the column and cycle half-life.

Genetic engineering has provided a cost effective method for the
production of large quantities of pharmaceutical proteins. The
next major challenge is the isolation of these proteins in a highly
purified form in high yield at low cost. The development of a cost
effective isolation procedure involves chosing methods for the
release of the protein from the cells, separation of the desired
protein from other soluble proteins and isolation of the protein in
a form required for stability and use.
    Except in cases where the protein is secreted by the cells,
the initial steps of cell concentration, cell disruption and
removal of cellular debris are common to all protein isolation
schemes, and will not be considered here.
    A major cost decision will be the selection of a method for
separation of the desired protein from other soluble proteins.
Classical methods for protein purification based on charge, size,
mass and solubility are individually non-specific and must be used
in appropriate combinations as experimentally determined. These
multi-step procedures increase capital costs, labor costs and the
time required for the purification. For labile proteins the
decreased yield of material due to degradation during the relatively long time required for purification is a major cost factor.
The use of molecular recognition as exemplified by the immunological complex formed by antigen and antibody, has definite advantages for protein purification. The application of chromatographic
methods using immobilized monoclonal antibodies can effect the
purification of a protein from a complex mixture in a single step
(1-6). Cost savings would result from reduced capital and labor
costs and increased isolation yields.

0097-6156/85/0271-0105$06.00/0

Criteria for Selecting an Immunosorbent

The industrial application of this technique requires the careful
selection of antibody, immobilization method and insoluble matrix.
The general requirements for an immunosorbent purification method
and the factors that affect these requirements are listed in Table
I.

Table I. Immunosorbent Requirements

REQUIREMENT	FACTORS AFFECTING REQUIREMENTS
Absorb protein from mixture	antibody, matrix, feed stock
Quanitative elution	protein, antibody, immobilization method
Adequate capacity	matrix, immobilization method
Retains capacity after repeated cycling	matrix, antibody, immobiliza-method, feed stock
Adequate flow rates	matrix, antibody

Only monoclonal antibodies allow one to select the specificity, affinity and stability of an antibody required for the
specific absorption of the protein, the quantitative elution of
the protein under conditions which retain its activity and the
capacity after repeated cycling. The immobilization method is
chosen to retain a high percentage of the antibody activity
affecting the capacity of the immunosorbent. Some methods of
immobilization form chemical bonds between the antibody and
matrix that are unstable in solutions that may be used for
loading or elution. The antibody can then bleed off the matrix,
thus affecting the purity of the protein and the capacity of the
immunosorbent after repeated cycling. A matrix is chosen for
its ability to support high flow rates, its lack of non-specific
absorption sites for other proteins in the solution, and
resistance to mechanical, proteolytic or microbial degradation.

Immunosorbent Capacity

After the appropriate monoclonal antibody, immobilization method
and matrix have been chosen according to the criteria discussed
above and methods previously described (7,8) the major factor in
determining the cost of this purification method is the amount
of antibody required. The amount of antibody required is determined by the capacity per cycle of the immunosorbent and the
number of cycles that can be utilized in a given process. The
capacity per cycle for the immunosorbent is given by Equation 1.

Equation 1. Immunosorbent Capacity per Cycle

$$C = A \times Y \times M.W.P/E.M.W.A. \times V \times e^{-0.693n/n\frac{1}{2}}$$

C =	Capacity/cycle (g)
A =	Total Antibody Immobilized (gL^{-1})
Y =	Immobilization Yield
M.W.P. =	Molecular Weight of Protein
E.M.W.A. =	Equivalent Molecular Weight of Antibody
V =	Column Volume (L)
n =	Cycle Number
$n\frac{1}{2}$ =	Number of Cycles Until Capacity = 0.5

As shown in Equation 1, the capacity per cycle is directly proportional to the amount of antibody immobilized, the immobilization
yield, the M.W. of the protein and the column volume and an
exponential function of the number of cycles. The amount of antibody immobilized will usually be less than 10 gL^{-1}. Higher activation of the matrix required for greater than 10 gL^{-1} loading
results in a decrease in the immobilization yield. The maximum
immobilization yield is 1.0 (100%) while 0.8 (80%) is not difficult
to obtain. The M.W. of the protein to be isolated is fixed. The
only way to increase the capacity per cycle significantly is to
increase the volume of the immunosorbent or increase the number of
cycles prior to reaching 50% of initial capacity (cycle half-life).
Increasing the volume of immunosorbent increases the amount of
monoclonal antibody required.

Since the cost of the antibody is the major cost, increasing
the volume of immunosorbent is the most expensive way to increase
capacity per cycle. The least expensive way to increase the
capacity of the isolation system is to increase the number of
cycles. The number of cycles that can be obtained in any given
purification is dependent on the cycle half-life and time-volume
constraints. The total amount of protein that can be isolated in a
given number of cycles is given in Equation 2.

$$\text{Total protein isolated} = Co \left[\frac{1 - e^{\frac{-0.693}{n\frac{1}{2}}n}}{1 - e^{\frac{-0.693}{n\frac{1}{2}}}} \right]$$

Where Co = capacity of 1st cycle
      n = number of cycles
      $n\frac{1}{2}$ = cycle half-life

In n = A x $n\frac{1}{2}$ cycles, where A is a constant, the total amount
purified is directly proportional to $n\frac{1}{2}$. For a given $n\frac{1}{2}$ the
total amount purified is an exponential function of n. The amount
of protein that can be purified with a given $n\frac{1}{2}$ in n number of
cycles is shown in Table II.

Table II.   Total Amount of Protein Purified as a Function of Cycle
Half-Life and Cycle Numbers

Cycle half-life $(n_{1/2})$	Total Protein Purified in n Cycles		
	$n = n_{1/2}$	$n = 2\ n_{1/2}$	$n = 3\ n_{1/2}$
5	3.86	5.79	6.76
10	7.46	11.20	13.07
20	14.68	22.02	25.69
40	29.11	43.66	50.94
80	57.96	86.95	101.44
160	115.67	173.52	202.44
320	231.10	346.66	404.45

Factors Affecting Cycle Half-Life

The cycle half-life is a function of the immobilization method, the
reagents used to dissociate the antigen/antibody complex, proteolytic and denaturing agents in the process stream and thermal
denaturation with time. We have immobilized the Fab' fragment of a
polyclonal goat anti-human IgG antibody as a model system.
Immunosorbent columns were prepared at 7 gL^{-1} and 1 gL^{-1} antibody
loading. Periodically, 50 mg of human IgG was loaded on the
columns in 10-50 mL of PBS or occasionally human serum. After
washing with 2 column volumes of PBS, they were eluted with 0.2 $\underline{M}$
acetic acid containing 0.15 $\underline{M}$ NaCl. The human IgG in the pass
through and eluant was routinely assayed by UV absorption. As a
check, a quantitative ELISA assay for human IgG was also used
periodically. Figure 1 shows the amount of human IgG bound as a
function of cycle number. Between cycles over a period of up to
195 days the immunosorbent was left on the bench at room temperature in PBS containing 0.2% sodium azide. For immunosorbent A,
(7 gL^{-1}) cycled 30 times over a 65 day period, $n_{1/2}$ = 19 cycles.
For B (1 mg ml^{-1}), cycled 50 times over a 195 day period,
$n_{1/2}$ = 53 cycles. Figure 2 shows the capacity of immunosorbent B
above cycled 40 times over a period of 3 days. In this case no
change in the capacity was observed as a function of cycle number.
This model showed that antibody loading and time between cycles had
a greater effect on capacity as a function of cycle number than
process stream or elution solvent. However, every system is not
expected to parallel this example.

Effect of Matrix on Cycle Time

A large cycle half-life and many cycles in the process will allow
the use of a lower immunosorbent volume (less antibody) if the
cycles are short enough. Under chromatographic conditions with the
proper "affinity selected" immobilized antibody, the formation of
the immunological complex is quite rapid. The limiting factor is
the volume that can be passed over the immunosorbent in a given
time. Assuming proper column design, the limiting factor is the
matrix on which the antibody is immobilized. Gels (agarose or

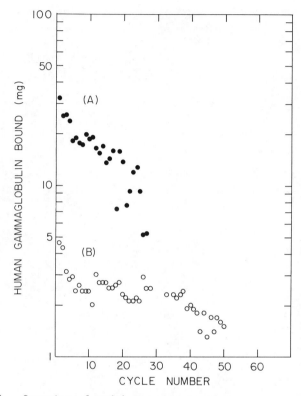

Figure 1.   Capacity of model immunosorbent systems as a function
of cycle number.   (A) Goat anti-human IgC Fab' fragment immobilized at 7 g/L.   Column volume = 6 ml.   (B) Same as above except
that the initial antibody loading was 1 gL^{-1} and the immunosorbent
volume was 4.2 ml.

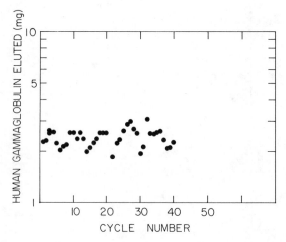

Figure 2. Capacity of a model immunosorbent system as a function
of cycle number.

acrylamide) are not suitable matrices due to flow restrictions. A rigid bead with large pores (or a macroreticulate bead) is required. We have used solid matrices which allow flow rates of 20-50 column volumes hr^{-1}.

Selecting Elution Solutions

After elution from the immunosorbent, the purified protein must usually be isolated from the elution buffer either as a dry powder, concentrated solution or a suspension. This step may be required for stability, packaging, or to make it suitable for administration to humans or animals. Here the selection of an antibody which will allow the dissociation of the antibody-antigen complex under appropriate conditions for the final processing step is important. If the final step is freeze-drying, a volatile elution buffer such as acetic acid or ammonium hydroxide would be appropriate.

Effect of Cycle Half-Life on Final Isolation Costs

As the capacity of the immunosorbent column decreases with increasing cycle number the concentration of the purified protein decreases as the volume in which it is eluted is a function of the column volume. If the concentration of eluted protein is 1 in the first cycle, the average concentration eluted in n = n/2 cycles = 0.75. The average concentration in the second cycle half-life = 0.375 and in the third cycle half-life = 0.1875. As a general rule if the cost of the final isolation is 1 for the first cycle half-life, it will be 1.33 for the second cycle half-life and 1.71 for the third cycle half-life. Regardless of the final process step, the decrease in concentration of eluted protein with cycle number will increase the final isolation costs and must be weighed against the cost of antibody needed to increase column volume and decrease the number of cycles.

An Example

Ultimately, the cost of immunosorbent isolation will depend on the entire process and must be evaluated against alternative processes. Consider, as an example, the costs and decisions involved in the purification of urokinase. One course of drug therapy consists of 33 mg of urokinase (4,000,000 CTA units). At the hospital pharmacy the drug costs for one course of treatment are currently $3,000 (9), or $91,000/gram. There are approximately 76,000 patients in the U.S. that could be treated with urokinase therapy each year requiring an annual production of approximately 2,500 g. We have selected a monoclonal antibody that has allowed the purification of urokinase from urine, tissue culture media, and bacterial culture media in a single step with 85% retention of urokinase activity (6). This monoclonal antibody was immobilized at 2 gL^{-1} with an immobilization yield of 0.8 and a cycle half number of 300 cycles. The urokinase capacity for the first cycle would be 1.2 gL^{-1} of immunosorbent.

From a solution of 5,000 L. of fermentation broth containing 6% E. coli, 440 g of urokinase can be solubilized assuming the urokinase is 1% of the total protein. The isolation consists of

absorption on the immobilized antibody, washing the immunosorbent
with 2 column volumes, eluting with two column volumes and
reequilibration with two column volumes. If the matrix allows a
flow rate of 40 column volumes per hour, 85% of the time will be
used for absorption. To isolate 2,618 g of active urokinase seven
fermentation runs are required. The purification must be completed
in 24 hrs. allowing 20 cycles per day and 140 cycles total.
Allowing for the decrease in activity with cycle number the total
amount of immunosorbent required is 21.4 L. At 2 gL^{-1} loading and
$200 g^{-1} for the monoclonal antibody and $200 L^{-1} for the matrix
and immobilization, the total cost of the immunosorbent is $12,840.
This amounts to $4.90 g^{-1} of urokinase. In this example the low
costs are obviously attractive.

Literature Cited

1. Secher, D.S; Burke, D.C.   Nature 1980, 285, 446-50.
2. Staehelin, T.; Hobbs, D.S.; Kung, H.; Lai, C-Y; Pestka, S.
   J. Biol. Chem. 1981, 256, 9750-4.
3. Hochkeppel, H-K; Menge, U.; Collins, J.; Eur. J. Biochem.
   1981, 118, 437-42.
4. Stallcup, K.C.; Springer, T.A.; Mescher, M.F.; J. Immunol.
   1981, 127, 923-30.
5. Stenman, U-H; Sutinen, M-L; Selander, R-K; Tontti, K.;
   Schroder, J.   J. Immunol. Meth. 1981, 46, 337-45.
6. Vetterlein, D.A.; Calton, G. J.   Thromb. Haemostas 1983, 49,
   24-7.
7. Calton, G.J.   In "Methods in Enzymology", Jacoby, W.; Ed.; 1984,
   104, 381-387.
8. Calton, G.J.   In "System Design for Industrial Scale Purification
   of High Value Proteins By Immunosorbent Chromatography." 5th
   Int'l Symposium of Affinity Chromatography; Ed. Academic Press:
   New York, 1984; pp. 383-392.
9. Stambaugh, R.L.; Alexander, M.R. Am. J. Hosp. Pharm. 1981,
   38, 817.

RECEIVED September 7, 1984

A Rational Approach to the Scale-up of Affinity Chromatography

F. H. ARNOLD, J. J. CHALMERS, M. S. SAUNDERS, M. S. CROUGHAN, H. W. BLANCH, and C. R. WILKE

Department of Chemical Engineering, University of California at Berkeley, Berkeley, CA 94720

Affinity chromatography has great potential for the
separation of high-value products from dilute fermentation broths; one of the drawbacks is the lack of information that can be applied to scale-up. Most affinity chromatography is in fact a rather special case of
a fixed-bed adsorption process: the equilibrium is highly favorable, and the breakthrough curve is likely to be
constant pattern. These characteristics greatly simpli-
fy the modeling. When presented in the proper nondi-
mensional form, the experimental breakthrough curves
from small, lab-scale columns can be used to predict
the performance of large affinity columns. This paper
discusses the breakthrough models applicable to affinity
chromatography in terms of the rate-limiting mass transfer mechanisms. We have compared the breakthrough data
from a model monoclonal antibody-antigen affinity column
to the predictions of two of these models. A rational
approach to the scale-up of affinity chromatograph is
presented.

The remarkable selectivities shown by numerous biological molecules
have been used to great advantage in recovery and purification by affinity chromatography. Monoclonal antibodies can be produced against
a wide variety of substances, making affinity chromatography applicable to the purification of practically any macromolecule from complex
mixtures. This technique is quickly becoming feasible for industrial
separations.

Janson and Hedman (1) recently published an excellent review of
large-scale chromatography. Many of the broad process design and
operation considerations are the same for affinity chromatography as
they are for ion exchange or gel filtration. Most chromatography
models, however, are based on the assumption of small feed pulses
with linear equilibria (such as the widely-used plate theories (2))
and are not directly useful for affinity separations. In this paper
we discuss and compare experimental results with two fixed-bed adsorption models that can be used to predict the performance of affinity columns. These two models differ only in the form of the rate-

0097-6156/85/0271-0113$06.00/0
© 1985 American Chemical Society

limiting mass transfer step—one case being diffusion in the pore liquid and the other a diffusion step in the particle phase. These models can be used to optimize a separation or to predict the performance of a large-scale column, based on data from laboratory-scale
experiments.

For systems with the favorable equilibria typical of affinity
chromatography, the breakthrough curve for an ideal operation is a
step function. The solute in the feed is taken up completely, up
until the time at which the bed is saturated. At this instant the
effluent solute concentration becomes equal to the feed concentration. The operation is then switched to wash and elution. This adsorption step is considered ideal because 100% of the bed capacity
has been used, and no feed has been wasted. The total uptake is
simply the equilibrium capacity of the affinity packing.

Just as in other types of chromatography, mass transfer, axial
dispersion, and deviations from perfect plug flow all act to spread
out the breakthrough curve. If the column is switched to wash at
a particular effluent concentration, c_{BT}, than a portion of the bed
capacity has not been used (Figure 1). If, on the other hand, the
adsorption step is continued until the entire bed is saturated, an
amount of solute equal to the area under the curved portion of the
breakthrough history is wasted. A longer adsorption cycle time is
needed to reach the full bed capacity.

An outline for developing a qualitative scale-up strategy has
been presented by Eveleigh (3). A quantitative study of process
optimization and the efficient use of resources requires an understanding of the behavior of the large-scale column. If the immobilized ligand (antibody) is scarce, operation may be optimized by
using the full bed capacity each cycle, going to higher breakthrough
concentrations. On the other hand, if the feed recovery is the most
important factor, the adsorption step may be terminated at small
breakthrough concentrations, with less than full use of the column
capacity. In order to predict the effects of changes in operating
conditions, one must know how each influences the shape of the
breakthrough curve.

Breakthrough models

An analytical expression for the breakthrough curve can be obtained
by solving the equations describing continuity of a sorbate species
in a fixed bed, the equilibrium relation between the solute and the
sorbate, and the rate of adsorption and mass transfer, with the appropriate initial and boundary conditions. The exact solution of
the complete set of equations is often impossible, but affinity
chromatography lends itself to several convenient simplifications,
with the result that analytical solutions are available. The notation used here is that of Vermeulen (4).

During the adsorption step, the binding of solute by the affinity ligand is often very tight, essentially irreversible. For Langmuir, or constant separation factor adsorption, the equilibrium relation is

$$\frac{q^*}{Q_m} = \frac{K_L\, c}{1 + K_L\, c}$$

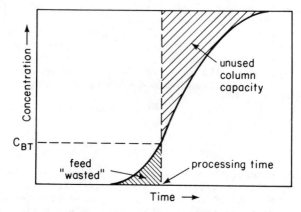

Figure 1. Breakthrough curve for affinity adsorption. If feed is
stopped at effluent concentration C_{BT}, a portion of the
column capacity has not been used. A small amount of
solute is lost in the effluent.

The separation factor R, defined by

$$R = 1/(1 + K_L c_{feed})$$

is analogous to the relative volatility in vapor-liquid equilibria. For favorable equilibria (R<1), the concentration profiles approach a constant shape as they progress down the column. This condition, known as constant pattern, is quickly established when R is very small, even in relatively short beds. The assumption of constant pattern greatly simplifies the mathematics. The form of the irreversible (R=0), constant-pattern breakthrough curve without axial dispersion depends on the rate-limiting mass transfer or reaction step. Glueckauf (5) has solved the problem for solid homogeneous particle diffusion with a linear driving force; Vermeulen (6) has proposed a solution based on a quadratic driving force that is closer to the exact solution. Hall (7) has solved the relevant equations for the breakthrough curve for pore diffusion, alone or combined with external film mass transfer. When the external film diffusion alone controls, the solution given by Michaels (8) can be used. The single mechanisms combined with axial dispersion have been investigated by Quilici (9).

In Vermeulen's notation, the breakthrough curve for the irreversible case can be presented in dimensionless form

$$X = f(N, T)$$

where X is the dimensionless concentration, T is the dimensionless time or throughput, and N is the number of transfer units in the column, defined in terms of the rate-limiting mass transfer or reaction mechanism. At constant pattern, the dimensionless solid concentration is equal to the liquid concentration, Y = X.

The number of transfer units for each mechanism can be estimated from known parameters and mass transfer correlations (4). For example, for a column with particles 0.01 cm in diameter, a superficial velocity of 0.01 cm/sec, and a solute bulk diffusivity of 7×10^{-7} cm^2/sec, the estimated number of transfer units in a packed bed of length L for the four mechanisms, axial dispersion, external fluid film mass transfer, pore diffusion, and solid homogeneous particle diffusion, are

$$N_d = \frac{Pe_p L}{d_p} \simeq 50 L$$

$$N_f = \left(\frac{k_f a_p}{u_o}\right) L \simeq 21 L$$

$$N_{pore} = \left(\frac{k_{pore} a_p}{u_o}\right) L \simeq 27 \left(\frac{D_{eff}}{D_{bulk}}\right) L$$

$$N_p = \psi_p \Gamma \left(\frac{k_p a_p}{u_o}\right) L \simeq 21\Gamma\left(\frac{D_p}{D_{bulk}}\right) L$$

The effective diffusivity of the solute in the particles can be several orders of magnitude smaller than the bulk liquid value.

Graham and Fook (10) studied the rate of protein adsorption in cellulosic ion exchange beads and found that it was limited by the rate
of diffusion into the particles. Their results were consistent with
an effective particle diffusivity 1/100 of the value in bulk solution. Thus N_{pore} and N_p can be significantly smaller than N_d and N_f.
Under these typical operating conditions, pore diffusion or solid
homogeneous diffusion would be the rate-limiting mechanisms. Axial
dispersion and film diffusion should have a small effect on the shape
of the breakthrough curve. For antigen-antibody and many other affinity systems, the adsorption reaction itself is often considered
to be very fast compared to the diffusion steps. This may not be
always the case, however, in the confined spaces of a pore.

The breakthrough curve for the irreversible case with pore diffusion is (7)

$$1) \quad N_{pore}(T-1) = 2.44 - 0.273\sqrt{1 - X}$$

For solid homogeneous diffusion (quadratic driving force), it is (6)

$$2) \quad N_p(T-1) = -1.69[\ln(1 - X^2) + 0.61]$$

Under some conditions there may be two rate-limiting mechanisms. The
solution to the combined pore and film diffusion problem (R=0) is (7)

$$3) \quad (T-1) = \left(\frac{1}{N_{pore}} + \frac{1}{N_f}\right) \frac{\left\{\phi(X) + \frac{N_{pore}}{N_f}(\ln X + 1)\right\}}{\frac{N_{pore}}{N_f} + 1}$$

where $\phi(X) \simeq 2.39 - 3.59\sqrt{1 - X}$

The breakthrough curve for solid homogeneous diffusion (linear driving force) combined with film mass transfer can be derived:

$$4) \quad N_f(T-1) = -mN_p(T-1) = \begin{cases} \ln\frac{X}{\beta} + 1 + m & 0 \leq X \leq \beta \\ \frac{(1-X)}{m \ln \frac{(1-X)}{(1-\beta)}} + 1 + m, & \beta \leq X \leq 1 \end{cases}$$

where $m = -N_f/N_p$ and $\beta = 1/(1 + N_f/N_p)$.

There may also be cases in which the equilibrium is not completely irreversible (i.e., very low solute concentration). A solution to the pore model for favorable equilibria (R < 1) is given by
Vermeulen (11). Values of X and $N_p(T-1)$ are tabulated in (7) for
various values of R < 1.

Experimental

To investigate the effects of the different mass transfer mechanisms,
breakthrough curves were generated on model monoclonal antibody affinity columns with two types of packings: Sepharose 4B (Pharmacia)
and controlled-pore glass (Electronucleonics, mean pore size 1273 Å).
Mouse monoclonal anti-benzenearsonate IgG was produced in this laboratory by batch culture in a 15 L fermentor. The IgG was purified

from the ammonium sulfate precipitate by affinity chromatography on
arsanilic acid-Sepharose 4B(12). The purified antibody was immobilized on CNBr-activated Sepharose 4B and on controlled-pore glass
with glutaraldehyde (13).
    The antigen for the feed was prepared by conjugating arsanilic
acid to bovine serum albumin (Sigma, RIA grade) by the diazotization
technique (14). As this material is deep red, flow distribution and
the progress of column saturation can be followed visually. Breakthrough curves were detected with an Altex 153 spectrophotometer at
335 nm.

Results

Typical breakthrough curves from the affinity column are shown in
dimensionless form in Figure 2 for the controlled-pore glass (CPG)
and in Figure 3 for the Sepharose gel. The curves generated on CPG
columns are in general better modelled by the pore diffusion equation; the curve shown in Figure 2 is best fit by Eq. 1, with N_{pore}
= 8. The breakthrough curves from the Sepharose column, on the
other hand, show a greater tendency to tail off at the upper end
and are better modelled by the solid diffusion equation. The best
fit of Eq. 2 to the data in Figure 3 gives N_p = 5.
    Using the following relation for $k_{pore}a_p$ (4), N_{pore} is consis-
tent with an effective diffusivity of 1.4×10^{-8} cm^2/sec.

$$\text{pore } a_p = \frac{60 \, D_{eff}(1 - \varepsilon)}{d_p^{\,2}}$$

If the bulk diffusivity of the conjugated BSA is close to the value
for unmodified BSA, 6.7×10^{-7} cm^2/sec, then $D_{eff}/D_{bulk} \simeq 1/50$.

Scale-up of Affinity Chromatography

The best way to illustrate the utility of these breakthrough models
in the scale-up of affinity chromatography is with an example. Let
us suppose we wish to predict the performance of a large column to
be packed with monoclonal-antibody-controlled-pore glass. The immobilization procedure has been optimized in separate experiments.
The breakthrough curves from a series of experiments on a 1.5 x 16.5
cm column indicate that pore diffusion is the rate-limiting mechanism and that N_{pore} = 8 at a superficial velocity of 0.011 cm/sec.
The larger column is to be 15 cm in diameter and 60 cm long.
    The number of transfer units for the large column can be cal-
culated from the definition of N_{pore} and the correlation for $k_{pore}a_p$.

$$N_{pore} = \frac{60 \, D_{eff}(1 - \varepsilon) \, L}{d_p^{\,2} \quad u_o}$$

The particle diameter d_p and the void fraction ε have not changed,
and the effective pore diffusivity should remain unaltered. Thus
the number of transfer units for the large column (LC) can be found
from

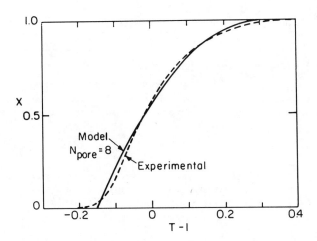

Figure 2. Breakthrough curve on 1.5 x 16.5 cm CPG–monoclonal anti-body column. $u_O = 0.011$ cm/sec, $\varepsilon = 0.4$, $d_p = 0.01$ cm, $T = 1$ at $V - \varepsilon v = 211$ ml ($\Gamma = 7.2$).

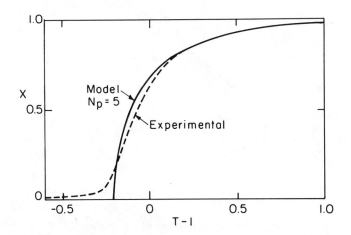

Figure 3. Breakthrough curve on 1.5 x 18.3 cm Sepharose 4B–mono-clonal antibody column. $u_O = 0.011$ cm/sec, $\varepsilon = 0.4$, $d_p = 0.01$ cm, $T = 1$ at $V - \varepsilon v = 117$ ml ($\Gamma = 3.6$).

$$\frac{N_{pore, LC}}{N_{pore}} = \frac{L_{LC}}{L} \times \frac{u_o}{u_{o,LC}} = \frac{60}{16.5} \times \frac{0.011}{u_{o, LC}}$$

or, $N_{pore, LC} = 0.32/u_{o,LC}$.

The dimensionless breakthrough curve is calculated using Eq. 1, and T can be converted back to volume throughput using the definition

$$T = \frac{V - \varepsilon v}{\Gamma v}$$

where $\Gamma = (V - \varepsilon v)/v$ is evaluated at $T = 1$ on the breakthrough curves from the small column. The predicted breakthrough curves for three different flow rates are shown in Figure 4.

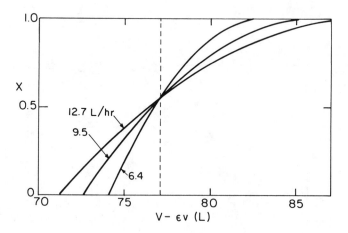

Figure 4. Breakthrough curves predicted by Eq. 1 for 15 x 60 cm column of CPG-monoclonal antibody. Equilibrium capacity is the solute equivalent of 77.1 L of feed ($v\Gamma$).

Conclusions

Affinity chromatography is a particularly simple example of fixed-bed adsorption: very tight binding of the solute during the adsorption step means that the shape of the breakthrough curve depends only on the rate-limiting mass transfer (or reaction) mechanism. Analytical expressions are available for a number of cases; four that can be useful in the scale-up of affinity chromatography have been presented here.

The procedure for predicting the performance of a large-scale column is straightforward when these dimensionless models are used. The rate-limiting mass transfer mechanism(s) can be identified in a series of experiments on a small column with the packing that is to be used in the large system. The parameters ka_p and Γ are obtained from the experimental data. New values of N are calculated from the definition for that particular mechanism and the operating conditions of the large-scale system. Once the number of transfer units is known, the dimensionless breakthrough curve can be generated with the appropriate breakthrough model, and the dimensioned curve can be obtained from the definitions of the dimensionless groups and the new column conditions. When the operating conditions are very different, it is a good idea to recalculate N values for the other mechanisms (axial dispersion, film mass transfer, etc.) to confirm that the correct model is being used.

Nomenclature

a_p external surface area of sorbent particles per unit packed volume (cm^{-1})

c solute concentration in liquid phase ($g\ cm^{-3}$)

d_p particle diameter (cm)

k^p mass transfer coefficient ($cm\ sec^{-1}$)

q average solid phase(sorbate)concentration ($g\ g^{-1}$particle)

u_o superficial velocity ($cm\ sec^{-1}$)

v^o column volume (cm^3)

D diffusivity ($cm^2\ sec^{-1}$)

L bed length (cm)

K_L Langmuir constant ($cm^3 g^{-1}$)

N number of transfer units (dimensionless)

Q_m maximum bed loading ($g\ g^{-1}$ particle)

R separation factor (dimensionless)

T throughput parameter (dimensionless) $= (V-\varepsilon v)/\Gamma v$

V throughput volume (cm^3)

X liquid phase concentration (dimensionless) $= c/c_{feed}$

Y solid-phase concentration(dimensionless) $= q/q^*_{feed}$

Pe_p packing Peclet number $\simeq 0.5$ for liquids in laminar flow

ε void fraction

ψ correction factor (= 0.590 for R = 0)

ρ_p bulk density of packing ($g\ cm^{-3}$ bed volume)

Γ distribution parameter $= \rho_b(q^*_{feed})/c_{feed}$ (dimensionless)

Literature Cited

1. Janson, J.-C.; Hedman, P. In "Advances in Biochemical Engineering"; Fiechter, A., Ed.; Springer-Verlag, Berlin, 1982; p. 43-99.
2. Van Deemter, J. J.; Zuiderweg, F. J.; Klinkenberg, A. <u>Chem. Eng. Sci.</u> 1956, 5, 271.
3. Eveleigh, J. W. In "Affinity Chromatography and Related Techniques"; Gribnau, T. C. J. et al, Ed.; Elsevier, Amsterdam, 1982; p. 293.

4. Vermeulen, T. In "Chemical Engineer's Handbook" Chapter 16;
    Perry, R. H.; Chilton, C. H. Eds.; McGraw-Hill, New York, 1973.
5. Glueckauf, E.; Coates, J, J. Chem. Soc. 1947, 1947, 1315.
6. Vermeulen, T. Ind. Eng. Chem. 1953, 45, 1664.
7. Hall, K. R.; Eagleton, L. C.; Acrivos, A.; Vermeulen, T.
    I & EC Fund. 1966, 5, 212.
8. Michaels, A. Ind.Eng. Chem. 1952, 44, 1922.
9. Quilici, R. E.   M.S. Thesis, University of California, Berkeley,
    1969.
10. Graham, E. E.; Fook, C. F. AIChE J. 1982, 28, 245.
11. Vermeulen, T.; Quilici, R. E. I & EC Fund. 1970, 9, 179.
12. Mishell, B.; Shiigi, S., Eds. "Selected Methods in Immunology";
    W. H. Freeman & Co., San Francisco, 1980; p. 281.
13. Weetall, H. H.; Hersh, L. S. Biochim.Biophys. Acta 1969, 185,
    464.
14. Tabachnick, M.; Sobotka, M. J. Biol. Chem. 1959, 234, 172.

RECEIVED September 10, 1984

Design of a New Affinity Adsorbent for Biochemical Product Recovery

HENRY Y. WANG and KEVIN SOBNOSKY

Department of Chemical Engineering, University of Michigan, Ann Arbor, MI 48109

Conventional biochemical product recovery processes generally contain a solid removal step in which cell debris and other solids are being removed from the product containing aqueous broth before further purification. Substantial amount of product loss occurs at this step due to binding and washing. A new affinity adsorbent has been developed by immobilizing small adsorbents in hydrogel beads. The internal mass transfer problems that exist in most commercial adsorbents can be reduced by using smaller size adsorbents. The adsorbent containing hydrogel beads can be quite large (1–3 mm) and can be easily recovered from the fermentation broth. Selective adsorption can also be achieved by changing the composition of the hydrogel and the types of the adsorbents. This concept has been shown to work well for several antibiotics and vitamins. The loading capacity of cycloheximide recovery has been shown to increase by 30% and the purity of the antibiotic eluted by organic solvents is above 90% which is significantly higher than the conventional extraction methods.

The process of recovering and purifying fermentation products in the biochemical industries is generally difficult and costly. The product can exist intracellularly or extracellularly and it may be sensitive to temperature change, extremes of pH, certain chemicals and enzyme activities of the microorganisms. Frequently, the energy and labor costs spent on recovery and purification of the fermentation product far exceed the cost of fermentation. This is especially true for intracellular recombinant protein products. The final concentration of the fermentation product is usually low (less than 0.2 wt%). Therefore, new processing techniques must be developed to improve the existing biochemical product recovery procedures. The conventional biochemical product recovery processes can be divided into four sections: 1. Removal of insolubles 2. Primary isolation of product 3. Purification

0097–6156/85/0271–0123$06.00/0

4. Final product isolation (1,2). At least one or more unit
operations are needed to accomplish each of these steps. Increase
in the number of recovery steps will definitely reduce the overall
extraction yield even though the product purity may be increased
(Figure 1). Certain amount of trade off will be needed in
achieving both the product yield and the product purity of a
specific product recovery scheme. In general, the drastic drop of
extraction yield occurred during the first two steps; namely
solids removal and primary isolation of product (Figure 1).

Whole beer processing using either solvent extraction or resin
adsorption have been reported to increase extraction yield (3,4).
This increase in recovery is because product losses due to binding
with solids and their subsequent removal can be eliminated. In
some cases, whole beer processing may eliminate the need for
filter aid in the initial step thus reducing cost or eliminate the
time consuming filtration step altogether. Whole broth processing
implies processing the fermentation broth without removal of the
insoluble fraction, this procedure therefore eliminates the
initial steps such as solid removal. Even though the major
advantage claimed, so far, has been to eliminate the solid removal
step, it is quite misleading because the solids still have to be
concentrated afterwards for waste disposal.

It is generally agreed among the practitioners of whole beer
processing that product removal in the presence of cells is more
difficult to attain, and requires that all physico-chemical
parameters of the operation such as fermentation medium
composition to be standardized. The purpose of this paper is to
describe a new approach to achieve whole broth processing using
immobilized solid phase adsorbents.

Adsorption allows the selective collection and concentration
onto solid surfaces of specific dissolved molecules from the
broth. Adsorption can be non-specific, for those mechanisms based
on polar, van der Waals and ionic interactions, or highly
selective for affinity binding based on biochemical means (1,8).
Commercially available adsorbants are generally granular porous
particles to provide extensive surface area, with void volumes
approaching 30-50% of the whole particle. Pore diameters are
usually less than 0.01 mm.

Many adsorbents, such as activated carbon and ion-exchange
resins, can efficiently separate antibiotics and other small
biologically active molecules from the fermentation broth.
Unfortunately, these adsorbents also interact with the microbial
cells and some of the dissolved nutrients. Thus, the use of
ion-exchange resins and activated carbon to remove fermentation
products is frequently associated with problems of simultaneous
removal of nutrients and side products. Substantial volume
reduction occurs but only limited purification can be achieved.
Commercial adsorbents and ion-exchange resins are available in
various matrices and sizes. Larger particles are preferred for
easy separation from the broth but they can be internal mass
transfer limited.

We have been interested in immobilizing different adsorbents
such as activated carbon powder and ion-exchange resins in
hydrogels such as calcium alginate or potassium carrageenan. The

purpose is to develop an affinity bead designed to increase both
mass transfer and selective purification for whole beer
processing. We are also interested in using these affinity beads
for in situ product removal.

Design of the Affinity Bead

One of the objectives of a biochemical separation system design is
to minimize the number of steps (Figure 1). One way of
accomplishing this is to perform the separation and concentration
in situ that is directly with the whole fermentation broth using
solids phase adsorbents. This type of separation requires the
design of an affinity bead that provides for selective product
removal from the fermentation broth.
   Various types of solid adsorbents have been used to
concentrate different biochemical products from fermentation
broths. The size of the solid adsorbent particle is important
because large macroscopic beads can easily be separated and
recovered from fermentation broths. However, large porous beads
exhibit internal diffusional resistance and depending on
processing time, all the binding sites of the adsorbent may not be
utilized, resulting in a lower adsorption capacity. Also, for
some adsorbents cell debris and proteinaceous materials may tend
to adhere to the surface of the solid adsorbent and would
contaminate the product in the subsequent elution process (7).
   We have been investigating the possiblity of immobilizing
solid adsorbents such as activated carbon, non-ionic polyermic
resins, monoclonal antibodies in small hydrogel beads (7), using
both K-carrageenan and calcium alginate to provide the hydrogel
matrix. Adsorbent concentrations in these hydrogel beads (Figure
2) can be as high as 30-35 wt %. Higher concentrations than 30 wt
% of the adsorbents weakening the bead structure, allowing
fragmentation under shear.
   These composite immobilized adsorbents (Figure 2) provide
additional selectivity for product absorption from the
fermentation broth. Very large macromolecules will be excluded
from the hydrogel and those that do penetrate will diffuse through
the gel at different rates depending on their size. If the binding
sites are selective in nature, only the desired product will be
adsorbed. The gel is reversible (with the presence or absence of
exogenous cations) and the adsorbents with the desired product can
be recovered from the gel matrices through washing and dissolving
the gel. Thus, both concentration (volume reduction) and
purification can be achieved. The large hydrogel beads can also
easily be recovered from the ferementation broth without
disrupting the microbial culture.

Adsorption of Cycloheximide to Immobilized Beads

Recovery of cycloheximide, a glutarimide, antifungal antibiotic
(M. W. 281) produced by Streptomyces griseus has been used as a
model system. The antibiotic fermentation was carried out
according to Kominek (5). The whole fermentation broth was
adjusted to pH 6.0 and stored in 4°C cold room. The antibiotic

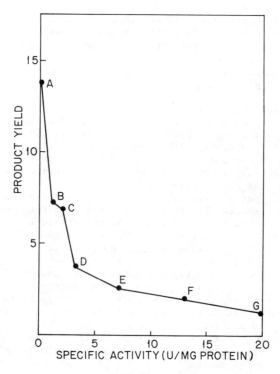

Figure 1. Correlation of extraction yield versus product purity
for a biochemical product recovery scheme. Key: A, after fermentation; B, after protein extraction; C, after concentration; D,
column 1; E, column 2; F, column 3; G, column 4.

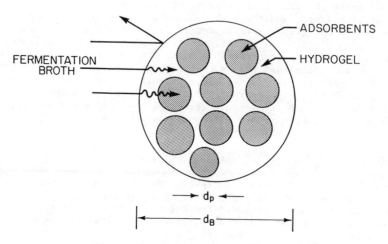

Figure 2. Design of an affinity bead.

has been shown to be stable for at least two weeks under such
condition.

XAD-4, a porous nonionic polymeric resin of polystyrene in
nature (Rohm and Haas, PA.) has been shown able to concentrate
cycloheximide from the fermentation broth (6,7). XAD-4 resins can
be immobilized into K-carrageenan or calcium alginate by adding
the resins (up to 30 wt%) in dissolved hydrogel solution and pump
the slurry through small tubing; droplets form by dripping into a
respective gelling solution. The resins are entrapped in the
hardened hydrogel beads. The size of the hydrogel beads can be
controlled by the size of the tubing. Smaller resin particles can
be obtained by pulverizing the resin particles by mechanical
means.

The adsorption kinetics of cycloheximide (CH) from either
aqueous solution or whole fermentation broth were carried out
using both plain XAD-4 resins and immobilized XAD-4 resins.

By determining the bulk concentration of CH (P), the batch
adsorption rate of CH using XAD-4 resin can be estimated. The
adsorption rate can be modeled as follows:

$$\frac{dQ}{dt} = K_s \ m \ (P-P^*) \tag{1}$$

Q is the loading capacity of CH by the polymeric resin and m is
the amount of the polymeric resin added. K_s is the specific
adsorption rate constant of CH and P^* is the surface
cycloheximide concentration which is in equilibrium to Q:

$$P^* = f(Q) \tag{2}$$

A Freundlich type adsorption isotherm can be used as long as P
be kept below 500 mg/l. As shown in Figure 3, the adsorption
isotherms of XAD-4 resin in both aqueous CH solution and
fermentation have been determined. In the fermentation broth, the
maximum adsorption capacity was reduced due to additional
impurities being adsorbed onto the resins. Close to 30% reduction
was observed. When immobilized XAD-4 resins in K-carrageenan are
used, similar adsorption patterns are observed. The maximum
loading capacity of 110 mg CH/gm resin can be achieved even in the
fermentation broth indicating exclusion of various impurities from
the XAD-4 resins which are immobilized inside the hydrogel.

By immobilizing the solid phase adsorbents such as XAD-4 resin
in the hydrogel, we actually add an additional diffusion
resistance to the adsorption. The product molecules need to
diffuse through the hydrogel beside diffusing through the stagnant
surface layer and penetrating through the resin particle to the
resin surface. As shown in Figure 4, the specific adsorption rate
constants of the free XAD-4 resin is much higher than the
immobilized XAD-4 resin. This is primarily due to the additional
diffusional resistance through the hydrogel. In both cases, the
specific rate constants decrease as the resins become saturated
with the product, cycloheximide. The specific adsorption rate of
the immobilized XAD-4 resins can be increased substantially when
pulverized XAD-4 resins (dp<0.15 mm) are used. The normal XAD-4
resins have an average diameter of 0.5 mm. By evenly distributing

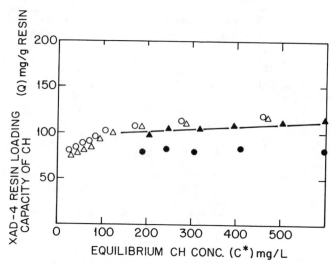

Figure 3. Adsorption isotherms of free XAD-4 resins and immobilized XAD-4 resins in K-carrageenan in aqueous CH solution and CH
fermentation broth. Key: O, aqueous, free resin; △, aqueous,
immobilized resin; ●, broth, free resin; ▲, broth, immobilized
resin.

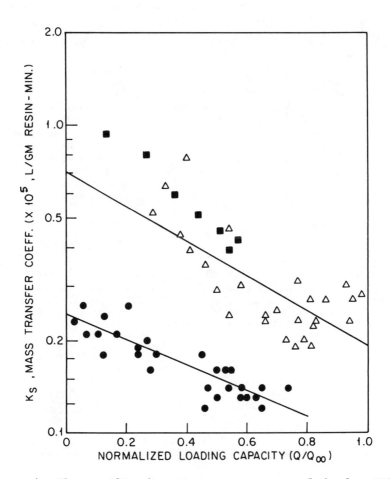

Figure 4.   The specific adsorption rate constants of the free XAD-4 resins and immobilized XAD-4 resins in K-carrageenan.   CH fermentation broth: ●, immobilized resin; △, free resin; ■, immobilized resin (dp < 0.15 min).

the smaller polymeric resins in the hydrogel beads, it is now
feasible to maintain the same adsorption capacity while increasing
the specific adsorption rates. Also, the immobilized hydrogel
beads can be recovered easily from the heterogenous liquid
mixture. This advantage has significant impact if the adsorbents
are very expensive. For example, monoclonal antibodies have been
suggested as tools to purify protein products. Most of the
monoclonal antibody based affinity adsorption only have a binding
efficiency of 15% or less. The higher costs of these affinity
ligands can be justified only if all ligands are being utilized to
bind the specific products and are recoverable.

   The immobilized XAD-4 resins in K-carrageenan and calcium
alginate are non-toxic to the microbial culture. Thus, the
adsorbents can be added directly to the fermenting broth. Both
fermentation and product separation can occur at the same time
(8). As shown in Table 1, the immobilized resin can be recovered
from the fermentation broth. The cycloheximide product can be
recovered from the immobilized beads by extraction using butyl
acetate. The immobilized beads maintained its mechanical strength
through the extraction. In some cases, the purity of the
cycloheximide in the solvent is close to 95-100% pure. Very
limited additional purification such as crystallization may be
used to make the final pure product.

Table I. Purity of Extracted Cycloheximide From the Free and
                 Immobilized XAD-4 Resins.

Fermentation	Solvents	Purity (%)
Shake Flask (Control)	Butyl Acetate	39
Fermentor (Control)	Butyl Acetate	24
XAD-4 Resin (Dispersed)	Butyl Acetate	54
Immobilized XAD-4-Resin	Butyl Acetate	87
Immobilized BACM Activated Carbon	Butyl Acetate	105

Conclusion

A novel approach to increase the overall extraction yield of a
microbial product has been developed. The primary isolation steps
of fermentation derived microbial products generally involve
procedures including solid removal, wash and volume reduction
using either solvent extraction or ion-exchange. These steps can
be further simplified by the use of whole broth extraction. In
the case of using solid phase adsorbents, extraction performance
can be drastically affected by the size and the nature of the
adsorbent used. If the adsorbent size is large, it exhibits
internal diffusional resistance (pore diffusion) and the resultant
mass transfer resistance reduces the overall rate of adsorption

and consequently, the final adsorption capacity due to competing
adsorption of impurities. On the other hand, separation of these
adsorbents directly from the fermentation beer is greatly enhanced
by the increased diameter of the immobilized adsorbents.
Obviously, an optimum size of the adsorbents where those opposing
effects need to be compromised. As shown, by immobilizing the
difficult-to-recover fine adsorbent particles in spherical
hydrogel beads such as calcium alginate, we can solve this
dilemma. The loading capacity of the adsorbent beads in the
hydrogel can be shown to be as high as 30 wt%. These new
immobilized adsorbent beads have much higher adsorption capacity
than the adsorbent alone. Additional advantages include easy
separation and recovery of the beads from the fermentation broth,
and partial purification due to selective adsorption of molecules
through the gel matrices.

Acknowledgments

The authors acknowledge Ms. Cynthia Miller for her editorial
assistance and typing of the manuscript. A research grant from
NSF (CPE-80-10868) made this work possible.

Literature Cited

1. Belter, P. A., 'Isolation of Fermentation Products',
    Microbial Technology, 2nd ed., 2, 403, 1979, Academic Press.

2. Edwards, V., 'The Recovery and Purifications of
    Biochemicals' Advances in Apl. Microbiol. 11, 159, 1969.

3. J. West and A. Patterson, 'Whole Broth Extraction'.
    Engineering foundation Conferences - Advances in
    Fermentation Recovery Process Technology, 1981.

4. Belter, P. A., Cunningham, F. L., and Chen, J. W.,
    'Development of a Recovery Process for Novobiocin:,
    Biotechnol. Bioeng. 15, 533m 1973.

5. Kominek, L. A. Antimicrob. Chemotherap. 7(6), 856, 1976.

6. Wang, H. Y., L. A. Kominek and J. L. Jost. Preceedings of VI
    International Fermentation Symposium 1, 60, 1981.

7. Wang, H. Y., G. F. Payne, and F. M. Robinson. Enhanced
    Cycloheximide Production through Neutral Adsorbant Addition.
    Presented at the XIII International Congress of Microbiology,
    Boston, Mass. 1982.

8. Wang, H. Y. 'Integrating Biochemical Seperation and
    Purification Steps in Fermentation Processes.' Biochemical
    Engineering III, Annals of the New York Academy of Sciences,
    413, 313, 1984.

RECEIVED October 5, 1984

Process Developments in the Isolation of Largomycin F-II, a Chromoprotein Antitumor Antibiotic

RAMESH C. PANDEY[1], CHABI C. KALITA[2], MARK E. GUSTAFSON[3], MARGARET C. KLINE, MICHAEL E. LEIDHECKER[4], and JOHN T. ROSS

National Cancer Institute, Frederick Cancer Research Facility, Fermentation Program, Frederick, MD 21701

Processes for the large scale isolation and purification of biologically active largomycin F-II, a chromoprotein antitumor antibiotic, produced by *Streptomyces pluricolorescens* MCRL-0367, from filtered fermentation broth and from mycelium (cell-paste) are described. Advantages of the more recent filtration/concentration units over the earlier methods, in process simplification, are discussed. These processes differ primarily in the initial steps used to recover and concentrate the material. In all cases, gel filtration on Sephadex G-100 and hydroxylapatite are used in the final stages of purification to yield gram quantities of purified biologically active largomycin F-II.

The isolation of largomycin F-II, a chromoprotein antitumor antibiotic, from the culture filtrate of *Streptomyces pluricolorescens* MCRL-0367 (1-3) was reported in 1970. It was shown to have biological activity against several tumors, including KB, P388, HeLa S-3 cells (0.1 µg/ml), Ehrlich ascites carcinoma, Sarcoma-180 and SN-36 leukemia (3). Some of its physicochemical and biological properties are summarized in Table I, and the partial amino acids sequence is shown in Figure 1 (4). Because of its biological activity, largomycin F-II was selected by the National Cancer Institute (NCI) for further formulation, toxicological and possible clinical studies, a situation requiring gram quantities of pharmaceutical grade material.

[1]Current address: Waksman Institute of Microbiology, Rutgers University, Piscataway, NJ 08854.
[2]Current address: 10919 Rawley Road, Mount Airy, MD 21771.
[3]Current address: Monsanto Company, St. Louis, MO 63167.
[4]Current address: 190 Stone Gate Drive, Frederick, MD 21701.

Table I. Physicochemical and Biological Properties of
Largomycin F-II

Producing organism (1)	:	*Streptomyces pluricolorescens* MCRL-0367
Nature (2)	:	Acidic, yellow amorphous powder
Molecular weight (4)	:	30,000, gel electrophoresis
		58,000, ultracentrifugation in PBS
		29,300, ultracentrifugation in 6M guanidine hydrochloride
Isoelectric point (4)	:	4.13 (pH gradiant 3.5-5.0)
UV λ max ($E_{1\ cm}^{1\%}$) (4)	:	
in 0.1 N HCl		272 (16.4), 420 nm (1.8)
in 0.1 N NaOH		288 sh (15.6), 525 nm (2.0)
Antibiotic activity (3)	:	Active against *S. aureus*
		S. luteau
Antitumor activity (3)	:	Active against KB, P388, HeLa, Ehrlich ascites, Sarcoma-180 and SN-36 leukemia

---

Preliminary scale-up assessments necessitated the investigation of the following criteria to achieve this goal:

1. Since titers of the antibiotic in the producing strain are generally low - strain and fermentation development work is necessary.
2. The development of assay methods is required since quantitative assay methods are not established.
3. Processes for the large scale isolation and purification are as yet not well established - therefore process development work is necessary.
4. As the material is destined for phase I clinical trials, it must conform to GMP, GLP and FDA regulations.

All four points have been addressed by various groups in the NCI-FCRF Fermentation Program. This paper will concern itself primarily with process development work for the purification of gram quantities of largomycin F-II. As work in one area impacts on the others, various aspects of all four points will also be discussed where appropriate.

Materials And Methods

General Comments. All high performance liquid chromatography (HPLC) analyses were carried out on a Waters Assoc. WISP 710B automatic injector and a Schoeffel SF770 Spectroflow variable-wavelength UV-Vis detector. The detector was set at 210 nm and 0.04 a.u.f.s. (cell volume 8 µl; path length 10 mm). Waters Assoc. protein analysis column I-125 (7.8 mm x 300 mm) and a mobil phase consisting of 0.05 $\underline{M}$ Na$_2$HPO$_4$/NaH$_2$PO$_4$, pH 6.8 at a flow rate of 1 ml/min and a chart speed of 0.5 cm/min was used for all the HPLC work.

Sephadex G-100 (particle size 40 ∿ 120 μ) purchased from Pharmacia Fine Chemicals, Division of Pharmacia, Inc., Piscataway, NJ 08854, was used for all Sephadex column chromatography. Diethylaminoethyl cellulose (DE-23) fibrous form, purchased from Whatman, Inc., Clifton, NJ 07014, was used for all DE-23 column chromatography.

Ultrafiltration and diafiltration were performed using Amicon DC-50 and DC-4 hollow fiber systems and/or a Millipore Pellicon unit, depending on scale of operation. A variety of pumps were used throughout the largomycin F-11 production campaign, including peristaltic, centrifugal, surge, as well as positive-displacement gear.

Largomycin F-II Assays. We have adopted the HPLC method of analysis for routine largomycin F-II assay, because it is simple, rapid (∿ 15 min/assay), quantitative, does not require extensive sample preparation and is selective in separating mixtures on the basis of molecular weight. However, this method cannot distinguish biologically inactive proteins from active proteins, nor could it resolve F-II from several proteases produced in the fermentation and of approximately the same molecular weight as F-II. Classical reverse phase gradient HPLC systems were also not useful, because the acidic nature of the solvent routinely used (0.05% TFA) resulted in inactivation of F-II. Due to these deficiencies HPLC assay is used in conjunction with two measures of biological activity, the Biochemical Induct Assay (BIA) (5,6) and *Micrococcus luteus* (ML) assays to follow the production and isolation of largomycin F-II.

Preparation of Samples for HPLC Assay.

From Broth. The broth sample is centrifuged and the supernatant filtered through a Millipore (0.45) filter. The filtrate is injected directly onto the HPLC column. A HPLC profile of such an injection is shown in Figure 2.

From Mycelium. Mycelia are suspended in phosphate buffer (0.05 M, pH 6.8, 5 ml/g) and sonicated for 5 min in a sonic water bath prior to centrifugation. After centrifugation the extract is processed in a manner similar to broth sample. A HPLC profile of mycelia extract is shown in Figure 2.

Biochemical Induct Assay (BIA) (5,6). The semi-quantitative spot test version (not the one-tube assay for the quantitative measurement of induction) was used for most of the work. The bacteria are poured in agar with or without rat liver S9 activation mix, onto large (24 cm x 24 cm) bioassay plates. Largomycin fermentation broth or test solutions are spotted onto the plates, allowed to incubate for three hours at 37°C, and overlaid with a second agar layer containing substrate. Within five or ten minutes, areas of induction are seen as red spots of insoluble dye formed by cleavage of the colorless substrate. Rapid sampling with the wide end of pasteur pipettes allows an operator to spot 100 samples on two plates in 20 minutes.

Micrococcus luteus (ATCC 9341) Assay. Mueller-Hinton agar plates were overlayed with 0.7% Difco Bacto-agar containing approximately 2.5 x 10⁶ cells of *M. luteus*/ml. When the agar is set, 15 μl of

1   5   10
Asp-Ile-Leu-Ile-Ala-Gly-Ala-Thr-Gly-Asn-

11   15   20
Val-Gly-Lys-Pro-Leu-Val-Glu-Gly-Leu-Leu-

21   25   30
Ala-Ala-Gly-Lys-Pro-Val-Arg-Ala-Leu-Thr-Arg-Asn---

---Ala-Ala-Leu-Phe-OH

Figure 1.   Proposed partial structure for apoprotein of largomycin
F-II. (4).

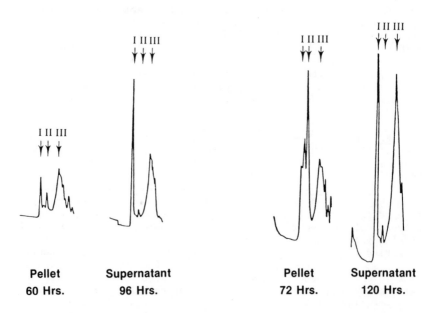

| Pellet | Supernatant | Pellet | Supernatant |
| 60 Hrs. | 96 Hrs. | 72 Hrs. | 120 Hrs. |

Figure 2.   High performance liquid chromatography profile of
centrifuged whole broth and mycelium extract.   Column: Waters
Assoc. I-125 protein analysis column (30 cm x 7.8 mm).   Mobile
phase: 0.05 M phosphate buffer, pH 6.9.   Flow rate: 1.0 ml/min.
Detector: UV at 210 nm.

sample or standard is spotted onto 6 mm paper discs, and the discs
are transferred wet onto the agar plate, and the plate is incubated
overnight at 37°C.
    After 20-24 hours, zones are measured in mm, and the log con-
centration of standard is plotted against zone size in mm. MIC are
determined by extrapolation of the standard curve to the Y-axis.
The concentration of largomycin F-II in samples is determined by
measuring zone size in mm and reading the concentration off the
standard curve.

Protease Assay. Nonspecific protease activity in largomycin F-II
was determined by Azocoll assay. Varying amounts of largomycin
F-II (20 µg ∿ 100/µg) dissolved in 20 mM Tris-HCl (pH 7.2) were
taken in stoppered tubes. The same buffer was added in each tube
to make the volume to 100 µl. To each tube 100 µl of azocoll
(100 mg/5 ml water) was added and incubated at 37°C for 1 hour.
The reaction was terminated by adding 2 ml of 5% trichloroacetic
acid to each tube. The tubes were centrifuged for 15 minutes at
25,000 rpm and the optical density of the supernatant was measured
at 366 nm. It was found that the measured protease activities were
linearly proportional to the amount of enzyme.

Results And Discussion

Isolation of Largomycin F-II. Figure 3 graphically illustrates HPLC
titers vs culture age for both cell free broth and mycelial extracts
for a typical 14-liter and 300-liter fermentations.
    Prior to attempting any process development work, it was
decided to evaluate the process reported by Yamaguchi *et al.* (2) for
the isolation of largomycin F-II.

Process of Yamaguchi *et al.* (2)

    In this process (Scheme 1) the largomycin complex was separated
into three components F-I, F-II and F-III by applying the following
procedure in succession: (A) gradient extraction with aqueous
ammonium sulfate solution, (B) isoelectric point precipitation, and
(C) size exclusion gel filtration on Sephadex G-100.
    This procedure yielded largomycin F-II that possessed little
biological activity, was impure on SDS polyacrylamide gels and contained protease activities. In addition the procedure resulted in
a low yield, resulting *ca.* 1.5 grams of material from a 2000 L fermentation.
    It has been observed and reported in the literature that
further purification of low specific activity F-II did not help in
increasing the biological activity of the material. At this stage,
then the following points were considered:

1. Largomycin F-II is only one of many "largomycins" of similar
    structure and bioactivity. Largomycin F-II is not necessarily
    the most biologically active of these components.
2. Largomycin F-II is a chromoprotein. Like the protein antitumor
    antibiotics neocarzinostatin (7-9) and macromomycin (10-12), the
    chromophore is probably responsible for the biological activity.

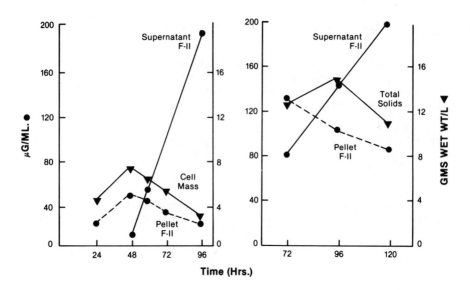

Figure 3.   Time course of largomycin F-II production in 14 liter
(left) and 300 liter (right) tanks.

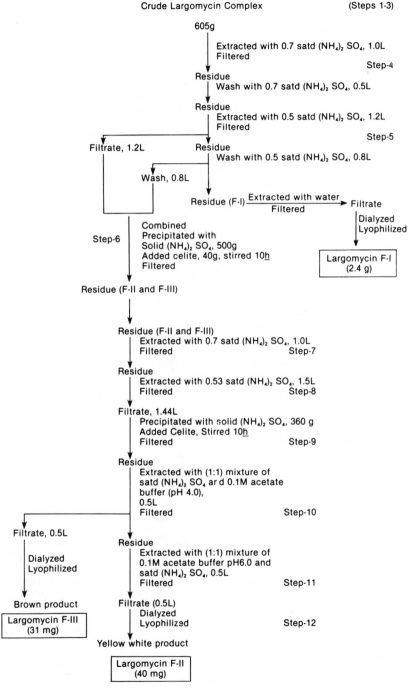

Scheme 1. Purification of largomycin F–II by the process of Yamaguchi et al. (2)

3. As a chromoprotein, largomycin is probably light sensitive.
Exposure to light results in an inactive protein with similar, if
not identical physicochemical properties.

4. During the fermentation, chromophore and protein are
probably produced separately, and only part of the chromophore is
attached to the largomycin F-II protein. The existence of free
chromophore explains the appearance of biological activity in a
fermentation broth before the appearance of largomycin F-II, and why
total bioactivity does not necessarily correlate with total amount
of F-II.

5. Several proteases are co-purified with largomycin F-II
which in solution degrade the largomycin F-II protein.

6. It has been possible to isolate the chromophore and the
protein under acidic pH with a suitable solvent.

All these observations are summarized in a non-chemical representation form in Figure 4. The lanterns with light represent the
active chromophore, the persons represent the protein, the dark lanterns represent the inactive chromophore. The effect of proteases
is shown by the breakage of the limb, head or body. The various
diagrams in a circle or box indicate what can be isolated by using
a suitable procedure in a particular step.

The primary drawback of the Yamaguchi, *et al.* process (Scheme 1)
was the large volume of the fermentation broth and the amount of
ammonium sulfate needed. Since a number of new technologies have
evolved in the protein purification during the last decade, it was
felt that a more efficient alternative process could be developed
to isolate and purify the biologically active largomycin F-II. With
these points in mind and the various equipment at hand, alternative
processes were developed for the purification of largomycin F-II
using the techniques of diafiltration, ultrafiltration, ion exchange
chromatography, gel permeation chromatography, hydroxylapatite
column chromatography and lyophilization.

NCI-FCRF Process I (Scheme 2). In this process filtered or centrifuged broth is used as the starting material for the isolation of
largomycin F-II. The filtrate was dialyzed against deionized water
to remove low molecular weight material, and concentrated in an
Amicon DC-50 hollow fiber ultrafiltration unit equipped with 10 K
cut off hollow fibers. After the initial ultrafiltration concentration, advantage is taken of the earlier work that largomycin F-II
stays in solution in 0.5 saturated ammonium sulfate solution, therefore an equal volume of saturated ammonium sulfate solution was
added to the concentrate and let stand at 4°C for 24 hours. [Con-
centration first reduces the amount of ammonium sulfate needed,
speeds the process, and makes precipitation and its recovery faster
and more efficient.] The concentrate is then filtered, dialyzed, and
concentrated through 10 K Amicon hollow fibers. Largomycin F-II was
then further purified by passage through Sephadex G-100 or DE-23
anion exchange cellulose columns. Both chromatographic procedures,
either alone or in conjunction with the other, gave material of
comparable quality. The largomycin F-II rich fractions, as judged
by HPLC and bioactivity were combined, dialyzed, concentrated and
lyophilized to a light yellow powder. The entire purification
scheme, together with the HPLC tracings of the material at various
stages, is shown in Scheme 2.

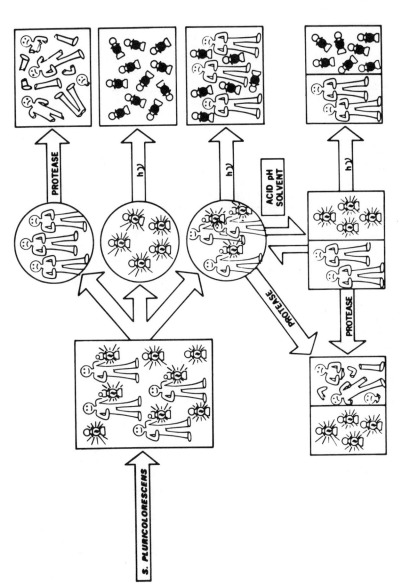

Figure 4.   Non-chemical representation of the production and isolation of the chromoprotein antitumor antibiotic, largomycin F-II.   The person with a lighted lantern is the biologically active form of largomycin F-II.

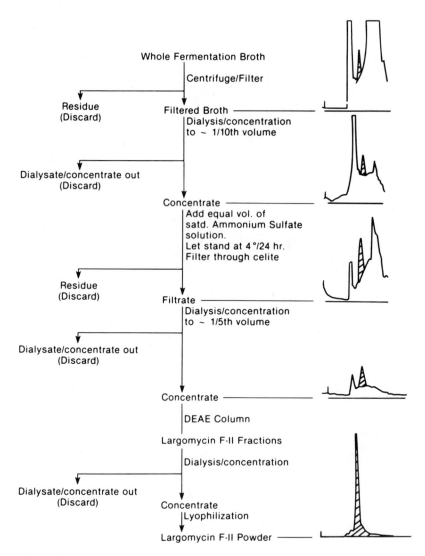

Scheme 2. Purification of largomycin F-II from filtered broth using recent methods (NCI-FCRF Process I). HPLC pattern of the largomycin purity is shown on the side.

The largomycin F-II isolated in this manner still contained
appreciable protease activity and showed impurity on SDS gels. It
was also noted at this time that if the fermentation is terminated
at about 48 hours, most of the largomycin F-II remains associated
with the mycelium (Figure 3). Material isolated from the mycelium
contained much less protease activity, and was also enriched in
F-II component (Figure 1) relative to supernatant. Also chromatography on hydroxylapatite as described by Montgomery *et al.* (4)
is effective in removing carbohydrate and protease from largomycin
F-II.

NCI-FCRF Process II-A (Scheme 3). In this process the starting
material for the isolation of largomycin F-II was the mycelium
(cell-paste). The cell-paste was suspended in potassium phosphate
buffer, and the suspension was subjected to cell rupture by freeze-
thaw steps and re-centrifuged or filtered. The crude largomycin
complex was dialysed and concentrated and the concentrate loaded on
to a Sephadex G-100 column. The column was eluted with phosphate
buffer. Largomycin F-II rich fractions were combined, dialyzed,
concentrated and lyophilized to a light yellow powder. The last
traces of carbohydrate impurities were removed by a hydroxylapatite
column. Based on biological assay and HPLC profile largomycin F-II
rich fractions were combined, dialyzed, concentrated and lyophilized
to pure largomycin F-II.
    Realizing the difficulty of freezing and thawing of large
quantities and the time taken in these operations another method was
developed for the extraction of largomycin F-II from mycelium. This
is summarized in Scheme 4. It was found that sonication or vigorous
stirring of the mycelia with deionized water or 0.01 $\underline{M}$ phosphate
buffer liberated most of the largomycin F-II in the first two
extractions (Scheme 4).

NCI-FCRF Process II-B (Scheme 5). In this process the freeze-thaw
step of mycelia suspension, to liberate largomycin F-II, of NCI-FCRF
Process II-A has been replaced by vigorously mixing the mycelia suspension for 1 hour. The results of a 240 gallon fermentation run
are summarized in Scheme 5. A comparison of the properties of the
isolated product with the standard compound is shown in Table II,
and their polyacrylamide gel electrophoresis pattern is shown in
Figure 5.
    One problem encountered during large scale purification of
largomycin F-II was the constant precipitation of aggregated materials, mainly impurities, during ultrafiltration. This material
could not be removed by centrifugation, or filtration through
celite. Microporous dead-ended filters quickly became clogged.
This high molecular weight material could be effectively removed
from the majority of the F-II by passage of the F-II through a 0.45
μ durapore membrane with excellent recovery and efficient removal
of high molecular weight proteins. These results are summarized in
Scheme 6 and Figure 6.
    The other aspect of largomycin F-II purification that was
routinely discussed within the Fermentation Program was whether it
was better to isolate F-II from the supernatant or from the myce-

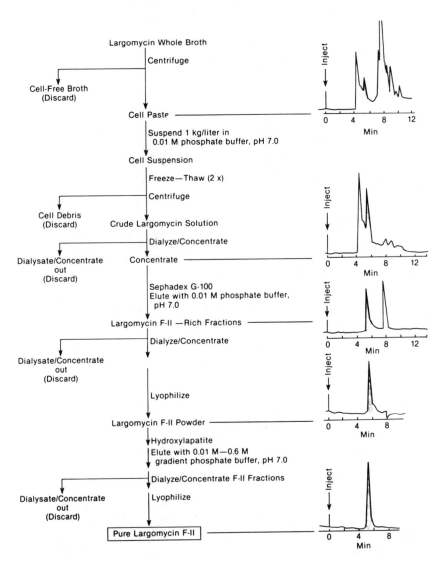

Scheme 3. Extraction and purification of largomycin F–II from cell-paste (NCI–FCRF Process II–A). HPLC pattern of the largomycin purity is shown on the side.

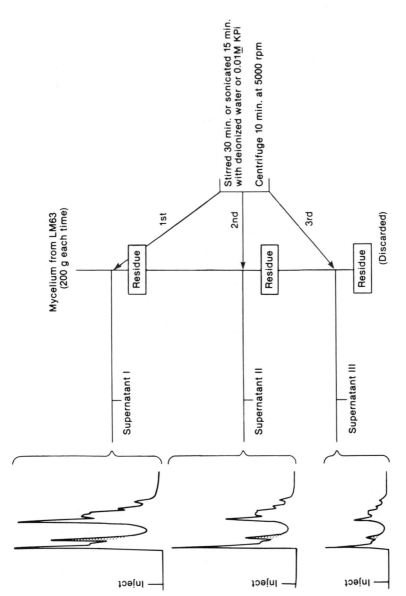

Scheme 4. General scheme for the extraction of largomycin complex from the mycelium. Shaded peak represents the amount of largomycin F-II in the complex.

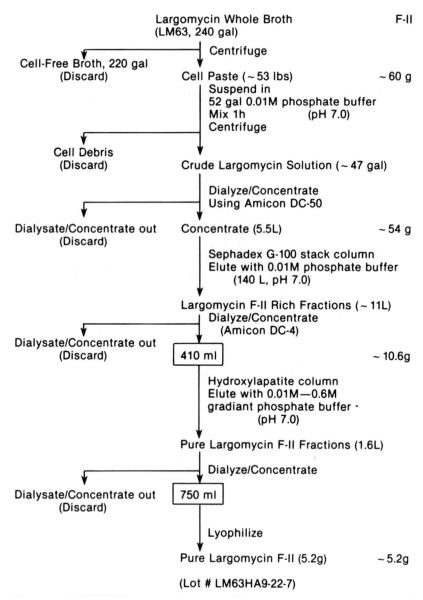

Scheme 5. Large scale purification of largomycin F–II from cell-paste using modern methods (NCI–FCRF Process II–B).

Properties	Frederick (Lot # LM63HA9-22-7)	Standard (RM-LM-F-II)
• Appearance	Flaky Yellow Solid	Granular Yellow Solid
• Solubility (0.1mM Phosphate buffer) pH 7.0	Soluble	Partially Insoluble
• Protein (%)	58	44
• Carbohydrate (%)	17.4	9.5
• MIC: *M. luteus* (μg/ml)	30	100
BIA (*E. Coli* strain 339)	0.6	5.0
• Protease (nmol/mg)	0.33	0.33
• HPLC (1mg/ml sol.)	Inject (50μl)	Inject (200μl)

Table II. Comparison of Physicochemical Properties of Isolated Largomycin F-II

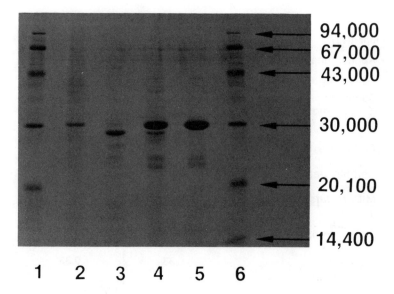

Figure 5. Polyacrylamide gel electrophoresis pattern of 1 and 6, markers; 2, crude largomycin from cell lysate; 3, purified largomycin F-II from filtered broth; 4, Sephadex G-100 purified largomycin F-II from cell lysate; 5, hydroxylapatite column purified largomycin F-II (final product from cell lysate).

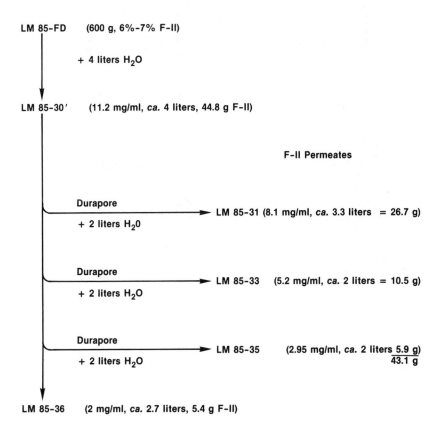

Scheme 6. Largomycin F-II Durapore processing scheme.

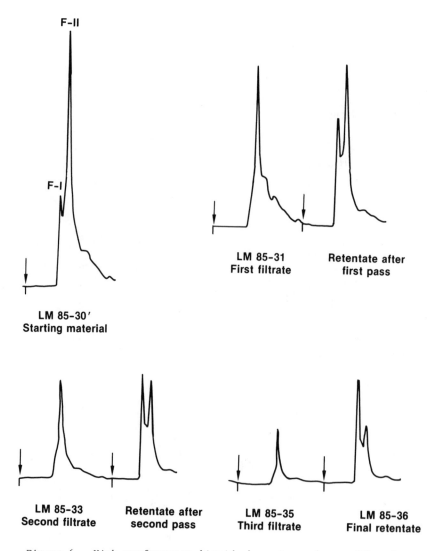

Figure 6.   High performance liquid chromatography profile of Durapore processing (condition similar to, as in Figure 1).

lium.   Crude material isolated from the mycelium was relatively
low in proteases and in high molecular weight impurities. Supernatant provided much more raw material initially (up to 300 µg/ml
by HPLC) but this F-II was high in protease activity and was con-
taminated with huge amounts of proteinaceous material from the
medium. The experiences learned in the development of process II,
namely the ability of hydroxylapatite to remove very similar impurities and the success of durapore membranes to effectively remove
large amounts of high molecular weight contaminants, spurred further
work on a supernatant process.

Fermentation development work was also progressing at this
time, and effective substitutes were found for the high molecular
weight components of the fermentation medium. Substitution of low
molecular weight nitrogen sources substantially reduced the amount
of high molecular weight protein impurities present in the supernatant.

Also about this time, we learned that largomycin F-II solutions
could be rapidly concentrated using a wiped film evaporator (WFE).
Reduced pressure low temperature evaporation is now routinely used
for the concentration of large volumes (500 gal.) of crude largomycin F-II solutions to a more manageable volume (50 gal.) prior to
ammonium sulfate precipitation or diafiltration procedures without
the loss of biological activity. Parameters can be adjusted so that
the material being concentrated is in contact with the heated evaporator rotar chamber for no more than a few seconds. The product
feed and the resulting concentrate emerging from the evaporator are
cooled.

NCI-FCRF Process III (Scheme 7). This is essentially modified
NCI-FCRF Process I, where advantage of all our experience and the
equipment at hand was taken to isolate high quality largomycin F-II
from filtered broth. The fermentation broth is clarified by filtration through celite in a plate and frame filter press, concentrated
ca. 5-fold with a WFE and centrifuged to remove solids. This rich
concentrate is dialyzed and concentrated using 10 K hollow fibers
on an Amicon DC-50. This dialyzed concentrate is treated with an
equal volume of saturated ammonium sulfate, filtered, and partially
dialyzed and concentrated before passage through Durapore cassettes.
The durapore permeate was concentrated and passed through a column
of Sephadex G-100. Largomycin F-II rich fractions are pooled
concentrated and lyophilized or immediately applied to a column of
hydroxylapatite which was eluted with 0.01 $\underline{M}$ phosphate buffer.
Largomycin F-II rich fractions from this flash HA column were
pooled, concentrated by ultrafiltration and carefully chromatographed on hydroxylapatite. Pure largomycin fractions from this
column were pooled, filtered, and lyophilized to give a light
yellow, highly hygroscopic powder that was equivalent in bioactivity
and purity to the cake derived material. This first attempt with
supernatant process produced *ca.* 15 g of high quality F-II from
about 400 gal. of fermentation broth. The procedure is depicted in
Scheme 7.

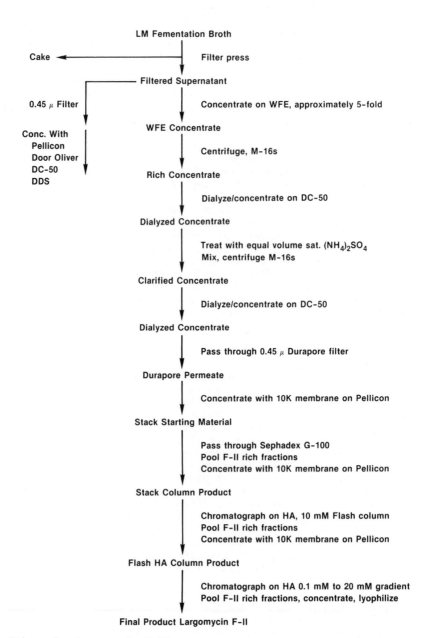

Scheme 7. Largomycin F-II supernatant recovery process (NCI-FCRF Process III).

Conclusions

Process development work on the purification of largomycin
F-II has resulted in two processes, one from cell-paste (NCI-FCRF
Process II-B) and the other from cell free broth (NCI-FCRF Process III). Both the processes have advantages over the Yamaguchi
et al. process in the number of steps, simplicity and yielding a
better biologically active product. Over 30 g of GMP quality,
homogeneous (on SDS gels) bioactive material has been produced from
approximately six successful mycelial recovery runs and two supernatant fermentations. Approximately 15 grams of these were derived
from 2-200 gallon supernatant runs, with an overall recovery of
between 5-10%. This material is currently undergoing formulation
and toxicity studies at the National Cancer Institute.

Acknowledgments

This research was sponsored by the Public Health Service,
National Cancer Institute under Contract No. NO-1-CO-75380 and
NO-1-CO-23910.
We thank Drs. John Douros and Matthew Suffness of the National
Cancer Institute and all the personnel of Fermentation Program,
NCI-FCRF, who were involved with this project at one time or the
other for all their help, interest and encouragement.

Literature Cited

1.   Yamaguchi, T.; Furumai, T.; Sato, M.; Okuda, T; Ishida, N.
     J. Antibiot. 1970, 23, 369-372.
2.   Yamaguchi, T.; Kashida, T.; Nawa, K.; Yajma, T.; Miyagishima,
     T.; Ito, Y.; Okuda, T.; Ishida, N.; Kumagai, K. J. Antibiot.
     1970, 23, 373-381.
3.   Yamaguchi, T.; Seto, M.; Oura, Y.; Arai, Y.; Enomoto, K.;
     Ishida, N.; Kumagai, K. J. Antibiot. 1970, 23, 382-387.
4.   Vandre, D.D.; Zaheer, A.; Squier, S.; Montgomery, R.
     Biochemistry 1982, 21, 5089-5096.
5.   Elespuru, R.K.; Yarmolinsky, M.B. Environ. Mutagen. 1979,
     1, 65-78.
6.   Elespuru, R.K.; White, R.J. Cancer Res. 1983, 43, 2819-2830.
7.   Ishida, N.; Miyazaki, K.; Kumagai, K.; Rikimaru, M.
     J. Antibiot. 1965, Ser. A. 18, 68-76.
8.   Meienhofer, J.; Maeda, H.; Glaser, C.B.; Czombos, J.; Kuromizu,
     K. Science 1972, 178, 875-876.
9.   Napier, M.A.; Holmquist, B.; Strydom, D.J.; Goldberg, I.H.
     Biochem. Biophys. Res. Commun. 1979, 89, 635-642.
10. Chimura, H.; Ishizuka, M.; Hamada, M.; Hori, S.; Kimura, K.;
     Iwanaga, J.; Takeuchi, T.; Umezawa, H. J. Antiobiot. 1968,
     21, 44-49.
11. Yamashita, T.; Naoi, N.; Watanabe, K.; Takeuchi, T.; Umezawa,
     H. J. Antiobiot. 1976, 29, 415-423.
12. Naoi, N.; Miwa, T.; Okazaki, T.; Watanabe, K.; Takeuchi, T.;
     Umezawa, H. J. Antibiot. 1982, 35, 806-813.

RECEIVED September 11, 1984

Approaches to Cephalosporin C Purification from Fermentation Broth

M. E. WILDFEUER

Eli Lilly and Company, Lafayette, IN 47902

Different antibiotics generally lend themselves to
specific purification techniques. Ceph C, however, is
fairly unique since its separation from broth can be
carried out by a variety of methods, each of which
illustrates a particular approach to antibiotic
isolation. Part of the reason for this versatile behavior results from the fact that ceph C is not the
end use product, and for this reason modifications
to the molecule to facilitate its separation or alter
its physical characteristics--without affecting the
active part of the ceph C molecule--are possible.

Examples of techniques used for purification of ceph
C or its derivatives include: (1) carbon adsorption,
(2) non-ionic resin adsorption, (3) ion exchange,
(4) solvent extraction of derivatives, (5) precipitation of derivative acid or salt, (6) precipitation
of metal salt, (7) broth drydown, (8) azeotropic extraction, and (9) enzymic modification. Many of these
methods are not merely laboratory curiosities but have
been successfully applied to large scale production of
ceph C.

The initial isolation of antibiotics and other fermentation products
from fermentation broths can take many routes. Those that have been
used on a large scale have included the following:
- Carbon adsorption
- Non-ionic resin adsorption
- Ion-exchange adsorption (anion or cation)
- Solvent extraction
- Precipitation
- Broth drydown
- Azeotropic extraction
- Enzymic modification

0097–6156/85/0271–0155$06.00/0
© 1985 American Chemical Society

Although one or the other of these techniques has typically
been applied to a particular antibiotic or class of antibiotics,
cephalosporin C is rather unique in that each of these methods can
be utilized successfully for its isolation.

To get some perspective on purification complexities, Table I
lists some of the major components found in a fermentation broth.

Table I. Broth Impurities

Classification	Examples
Insoluble Components	Mycelia
	Non-utilized raw materials
	Insoluble salts
	Immiscible oils
	Insoluble metabolites
Soluble Components	Soluble non-utilized raw materials
	Soluble metabolites
	Polysaccharides and sugars
	Proteins and amino acids
	Lipids
	Nucleic acids
	"Unique" metabolites

The mycelia and other insoluble substances are usually removed
by a filtration step, although whole broth resin or solvent extraction methods may be used. It should be noted that cephalosporin C is soluble in fermentation broth (some antibiotics are not).
The magnitude of the isolation challenge is illustrated by the fact
that 20-70 million liters of broth may typically be harvested
annually. Systems capable of handling broth flow rates of 100-200
liters per minute might be required.

The structure of cephalosporin C, a β-lactam antibiotic, is
shown in Figure 1. Under neutral, but especially basic conditions,
it is hydrolyzed to desacetyl cephalosporin C. In acid cephalosporin
C lactone is formed (1,2) (both of these are shown in Figure 1). In
order to minimize these degradations, it is important that cephalosporin C broth be processed rapidly, avoiding extremes of pH and
keeping temperatures low.

Activated Carbon

Activated carbon will effectively remove cephalosporin C from broth
(3,4); elution is effected with dilute aqueous solvents. The carbon
column eluate may then be purified further by adsorption and elution
of cephalosporin C using an anion exchange resin, since most of the
competing strong anions are not adsorbed to carbon. An example of a
carbon-anion exchange route is seen in Figure 2.

One of the difficulties of this method is the requirement that
carbon be periodically reactivated or replaced. Also, as mentioned,
elution and regeneration require solvent, and capacity for cephalosporin C is relatively low.

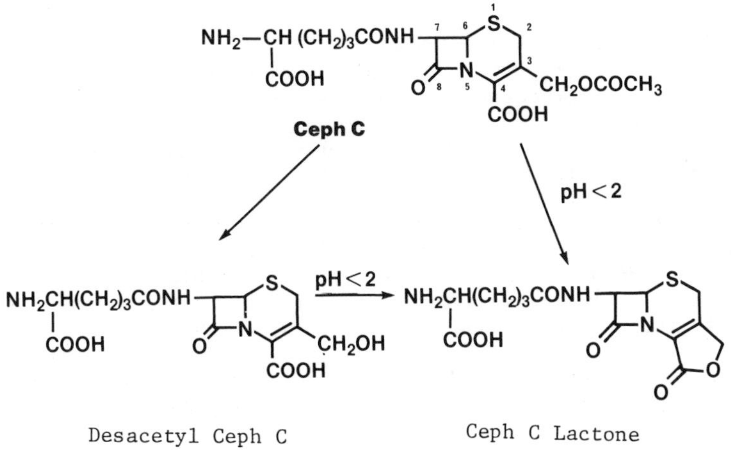

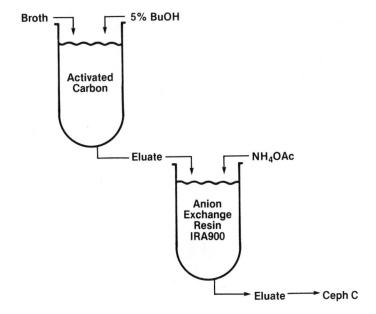

Figure 1.   Mild degradation of cephalosporin C.

Figure 2.   Carbon adsorption of cephalosporin C.

Non-Ionic Resin

A second adsorption method which is similar to activated carbon is
shown schematically in Figure 3. This involves the use of a non-
ionic polymeric adsorbent (5-8), such as Amberlite XAD2 or Diaion
HP20.
    Unlike carbon, these resins are very stable and can be reused
for hundreds of cycles. Another advantage is their high selectivity,
preferentially adsorbing cephalosporin C over desacetyl cephalosporin C. As with carbon, elution is usually carried out with
aqueous solvents and the resulting eluate, free of strong anions, is
adsorbed onto an anion exchange resin column and eluted with a
neutral salt. The process is rather independent of the anion-cation
makeup of broth, and--because of its high selectivity--gives a
high purity product. It again requires the use of solvent for
elution and regeneration and has a fairly low capacity for cephalosporin C, although resin manufacturers are developing more efficient
resins for this application (9,10).

Anion Exchange

Although pure cephalosporin C is readily adsorbed by strong or weak
base anion exchange resins, its isolation from broth is complicated
by the presence of numerous other anions. As illustrated in Figure 4,
it is necessary to have several columns in series, with a lead anion
exchange column used to adsorb stronger anions and a trail anion
exchange column to adsorb cephalosporin C (11-14). A cation exchange resin column in the hydrogen form (either before or after the
lead anion exchange column) is used to lower the pH without adding
anions to the broth, and helps to make the system more selective for
cephalosporin C adsorption. Alternatively, a single cation-anion
exchange mixed bed column has also been proposed ahead of the trail
anion exchange column. The adsorbed cephalosporin C, along with
other intermediate strength anions, are eluted with neutral salts
such as potassium acetate.
    Obstacles to the successful commercial application of anion
exchange for initial cephalosporin C isolation include the requirement for large volume lead resin columns to remove strong anions,
the need to adequately balance the lead and trail anion exchange
columns to minimize cephalosporin C losses on the lead columns, restrictions placed on fermentation media since strong anion salts
need to be kept to a low level, and the coadsorption of other intermediate strength anion impurities onto the trail resin.

Cation Exchange

Since it is a zwitterion, cephalosporin C can also be adsorbed on low
cross-linked sulfonic acid cation exchange resins, as shown in
Figure 5. Although used successfully for cephamycin C (15), a related
β-lactam antibiotic, the strong acidity of these resins results in
substantial degradation of cephalosporin C to its lactone. (Cephamycin C contains a carbamoyl ester in place of the acetyl ester at
C-3, and is apparently more stable.) Elution of the adsorbed
cephalosporin C with acetate buffers yields eluates with high lactone
content.

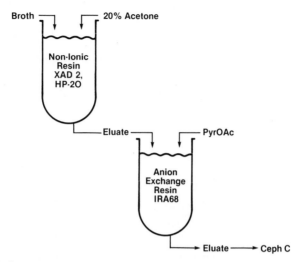

Figure 3.   Non—ionic resin adsorption of cephalosporin C.

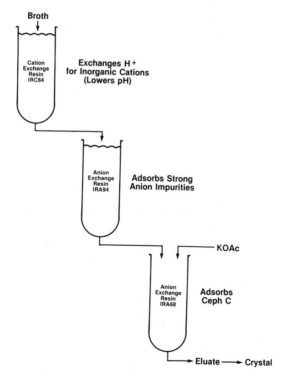

Figure 4.   Anion exchange adsorption of cephalosporin C.

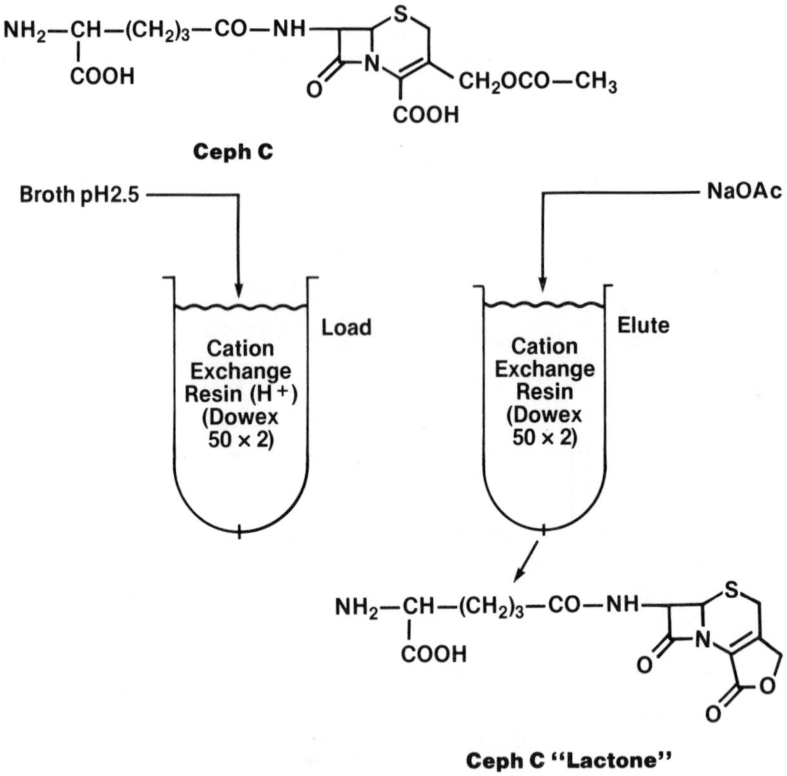

Figure 5.   Cation exchange adsorption of cephalosporin C.

Final Cephalosporins

A number of approaches to recovery of cephalosporin C take cognizance
of the fact that cephalosporin C is merely an intermediate in the
synthesis of cephalosporin antibiotics. Figure 6 shows those antibiotics currently manufactured at Lilly which are derived from
cephalosporin C.
    Substitution of the 7-amino side chain is accomplished by non-
aqueous cleavage of the α-aminoadipyl group to yield 7-aminocephalos-
poranic acid (7ACA). Reacylation at the 7-amino site by the desired
substituent is carried out in a subsequent step. By contrast, substitution at C-3 is usually accomplished by a direct displacement of
the acetoxy by the new group. Obviously, neither desacetyl cephalosporin C nor cephalosporin C lactone can participate in that substitution reaction.

N-Substitution

Substitution of the amine on the α-aminoadipyl side chain of cephalosporin C (Figure 7), using many of the derivatization methods
borrowed from classical peptide and amino acid chemistry, sufficiently alters the properties of cephalosporin C so as to, depending on
the derivative, make it solvent extractable or insoluble in aqueous
solutions. Figure 8 shows examples of many such derivatives which
have been prepared and which have been reported to aid in the isolation of cephalosporin C (16-40). The utility of this approach lies
in the fact that each of these cephalosporin C derivatives are able
to be cleaved to 7ACA in yields equivalent to (and sometimes better
than) yields achieved with cephalosporin C itself.
    Several points need to be considered in exploring this route:
(1) The reaction with the amine is not specific, and generally a
2-10 fold excess of reagent could be required for broth. (2) A water
miscible solvent is frequently needed for reaction optimization. (3)
The reaction may take several hours for completion. (4) The optimum
pH (8-10) and temperature (15-30°C) for the reaction can lead to
ceph C degradation. (5) The derivatizing reagent can be expensive.
(6) The enhanced solvent solubility of the derivative is frequently
beneficial for the cleavage step.

Extraction

Most of the derivatives shown in Figure 8 are solvent extractable at
low pH, and thus one of the classical methods used for antibiotic
purification becomes accessible to cephalosporin C. To be commercially feasible, solvents should be selective and only slightly miscible with water. Extraction efficiency should be sufficiently high
that multiple extractions are not required, and ideally should be
efficient at low ratios so as to effect a concentration of the
desired component. Emulsions and insoluble solids are anathema to
extraction. Using these criteria, extraction of most of the cephalosporin C derivatives at low pH are far from ideal since mostly
non-selective solvents (such as n-butanol and ethyl acetate) usually
work best; several extractions seem to be required, and derivatized
cephalosporin C broth upon acidification will frequently result in
emulsion formation. However, some derivatives behave better than

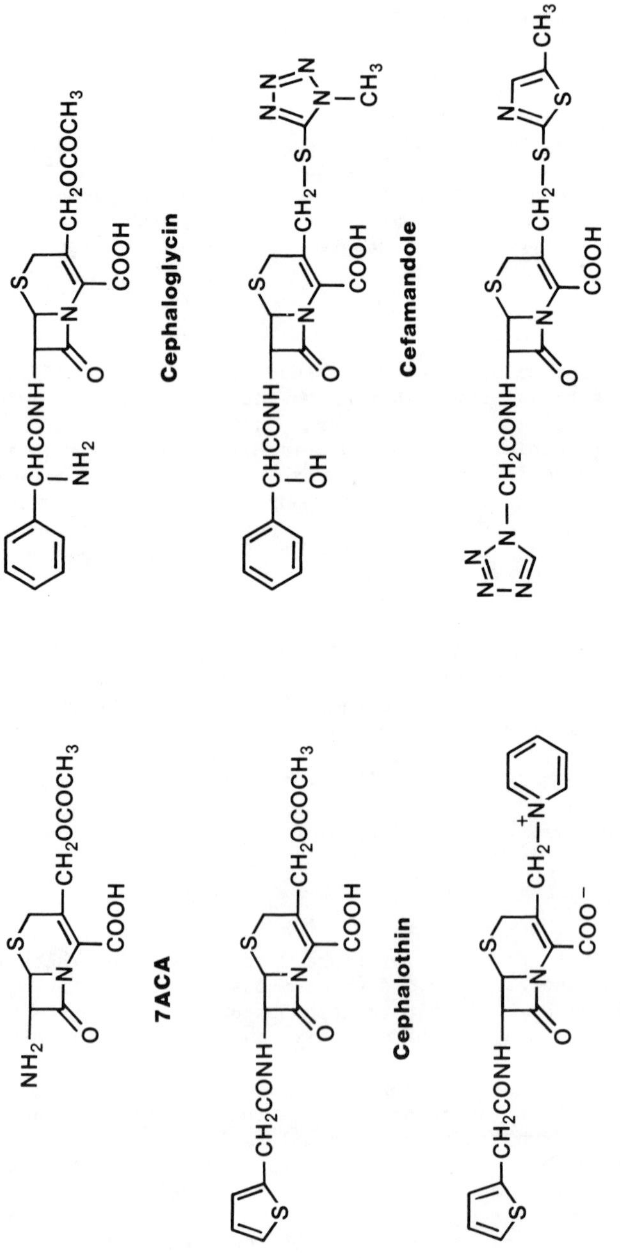

Figure 6. Examples of antibiotics derived from cephalosporin C.

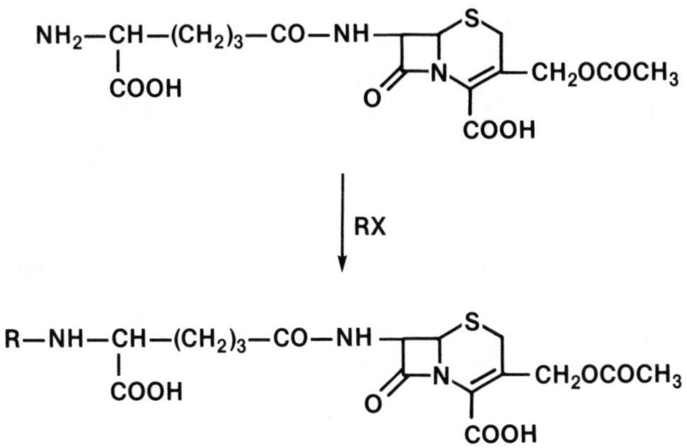

Figure 7. N-substitution of cephalosporin C. Many derivatives solvent extractable as acid. Some derivatives crystallize as acid.

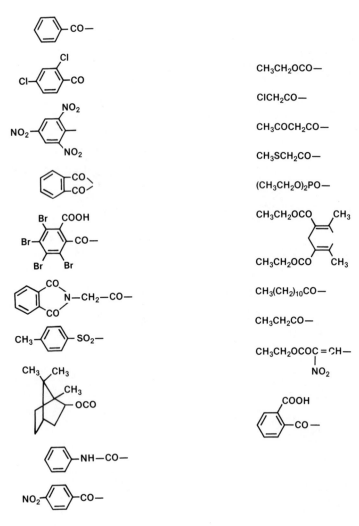

Figure 8.   Examples of N–substitution derivatives used in cephalosporin C purification.

others, and some of the difficulties can be alleviated by removal
of polymeric impurities by precipitation or ultrafiltration of broth
prior to extraction.

A noteworthy variation on the extraction approach is a procedure
proposed by Glaxo called "Extractive Esterification" (41,42). The
scheme, shown in Figure 9, reacts an N-derivatized cephalosporin C
with diphenyl-diazomethane, which combines with carboxyl groups to
form diphenylmethyl esters. This reaction is carried out in a water-
dichloromethane system, with the newly esterified cephalosporin
transferring from the aqueous phase to the organic phase. What is
especially attractive about this approach is the fact that dichloromethane is an ideal solvent for carrying out the cleavage of the
α-aminoadipyl side chain, and it therefore becomes unnecessary to
isolate the cephalosporin C derivative prior to cleavage. The
cleavage product is the diphenylmethyl ester of 7ACA, and an
additional hydrolysis step is needed.

As with other extraction procedures, emulsion and solids
problems need to be circumvented if starting at the broth stage.
Reagent synthesis and costs also need to be considered.

Precipitation

Precipitation of cephalosporin C can be accomplished by any of
several routes. These include:

- Crystallization of the potassium or sodium salt from purified
  aqueous solutions of cephalosporin C by concentration and/or
  addition of large volumes of a miscible solvent (13,36).
- The zinc salt (also copper, nickel, lead, cadmium, cobalt, iron
  and manganese) of cephalosporin C can be crystallized from
  purified aqueous solutions--some solvent is required (43).
- Insoluble derivatives such as the N-2,4,dichlorobenzoyl
  cephalosporin C and tetrabromocarboxybenzoyl cephalosporin C
  are crystallized as the acid from aqueous solutions (18,19,23).
- Organic base salts of derivatives (quinoline (32,34,37,44)
  dicyclohexylamine (45), dimethylbenzylamine (29), triethylene-
  diamine (19)) can be precipitated from solvent extracts or
  purified aqueous solutions.
- Sodium-2-ethyl hexanoate will precipitate the sodium salt of
  N-derivatized cephalosporin C from solvents (29,30).

Most of these precipitation methods will not work effectively
on broth, and hence cannot be used as an initial isolation step.

Drydown or Azeotropic Distillation

Although of itself broth drydown is not a very significant purification step, it can offer certain advantages. As the example in
Figure 10 shows, the dry product can be reslurried in another solvent
(in this case acetic acid), the cephalosporin C derivatized (with
acetic anhydride) and finally isolated as the zinc salt of N-acetyl
cephalosporin C (46).

In a related method, also shown in Figure 10, water can be
removed via an azeotrope with another solvent. In the example used,
broth derivatized with chloroacetyl chloride is added to cyclohexanone; after vacuum distillation of the water the cyclohexanone contains insoluble solids as well as dissolved N-chloracetyl cephalos-

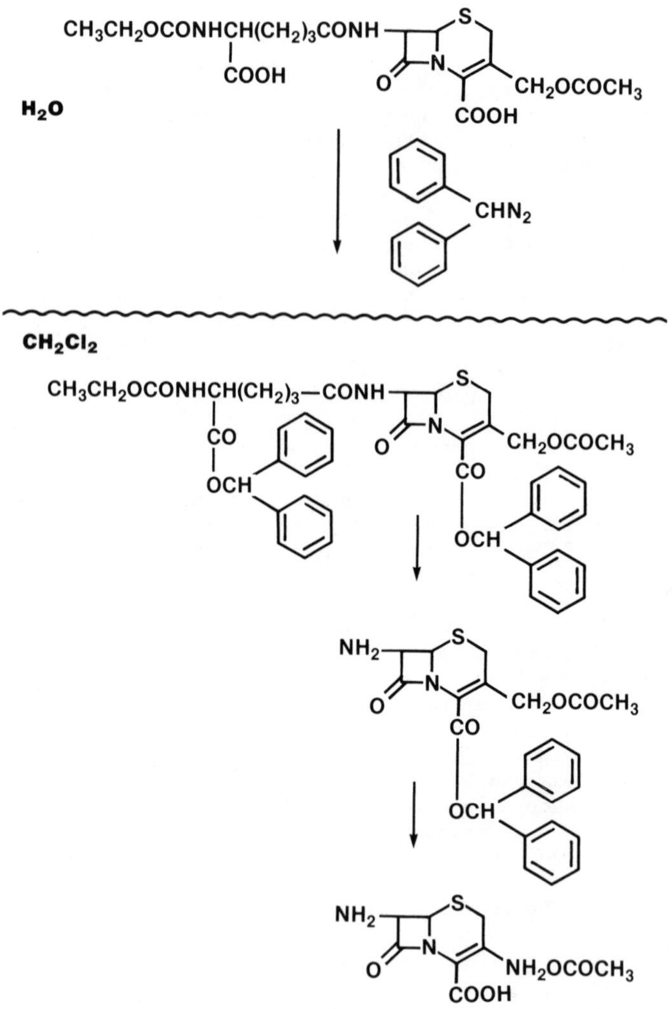

Figure 9.   Extractive esterification of cephalosporin C.

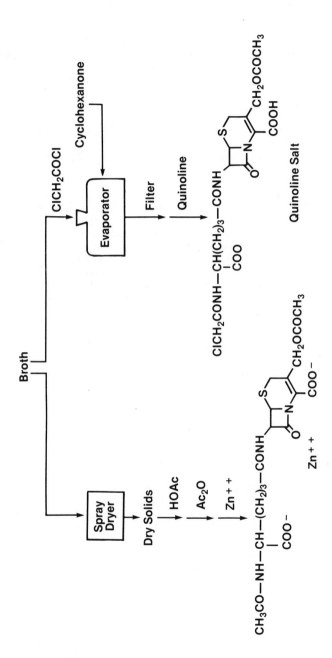

Figure 10. Broth drydown and azeotropic extraction for cephalosporin C isolation.

porin C. Following filtration of the solids, the quinoline salt
of the derivatized cephalosporin C is precipitated (47).
     Both of these methods require minimal waste treatment facilities;
however, evaporation requirements are substantial and initially, as
mentioned previously, these are concentration steps only. In addition,
cephalosporin C degradation can occur during the evaporation or
distillation steps. One major process advantage is that emulsion
problems are avoided since there is no solvent partitioning involved.

Acetoxy Displacement

As discussed earlier, cephalosporin C is an intermediate for other
cephalosporin antibiotics. Although some of these retain the C'-3
acetoxy group, for others the acetoxy is displaced by a thiol or
pyridinium compound. Where this displacement is generally carried
out at the 7ACA stage or beyond, it can be run earlier. As an
example, Figure 11 illustrates the reaction of 1-methyltetrazole-5-
thiol, used in cefamandole synthesis, with cephalosporin C.
     Acetoxy displacement in broth prior to isolation of cephalos-
porin C (6,21,31,37,48-50) is compared schematically in Figure 12
with the more usual approach to cephalosporin C purification.
Derivatization at C-3 can be used in conjunction with other methods
such as non-ionic resin adsorption or N-derivatized extraction.
However, no advantage in terms of reduced solubility or enhanced
extractability is indicated for these derivatives. This alternative,
however, does effectively eliminate one step in the chain and could
result in an overall yield benefit.
     Displacement of the C'-3 acetoxy, unlike N-derivatization, is
fairly specific and would not require a large excess of reagent.
However, it does limit flexibility and, if more than one derivative
is required, alternate isolation schemes may be needed. Also, the
reaction is somewhat slow and is carried out at relatively high
temperatures.

Enzymatic or Microbial Conversions

An innovative approach to cephalosporin C and 7ACA recovery involves
the enzymatic conversion of cephalosporin C to solvent extractable
intermediates and ultimately to 7ACA. The first step involves microbial deamination or chemical transamination to the α-ketoadipyl or,
after spontaneous decarboxylation, glutaryl cephalosporin C derivatives (51-54). These derivatives then--unlike cephalosporin C itself-
-can be enzymatically cleaved to 7ACA (55-60). These reactions are
shown in Figure 13.
     The deamination step has been carried out in broth, followed
by extraction of the deaminated derivatives (51). Immobilized
enzyme systems have been used for the 7ACA step (55).

Utilization of Desacetyl Cephalosporin C

No discussion of cephalosporin C isolation would be complete without
some comment on desacetyl cephalosporin C. This compound, always
present in significant quantity in broth, besides being a neutral
and high pH degradation product of cephalosporin C (2,61), is also
produced enzymatically by endogenous or contaminant esterases.

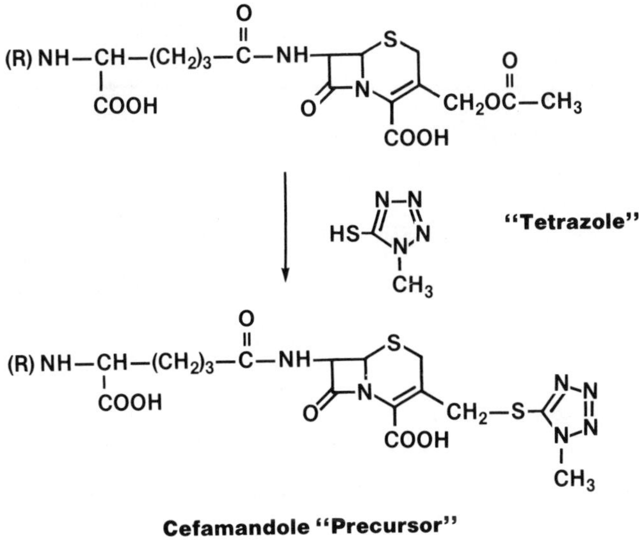

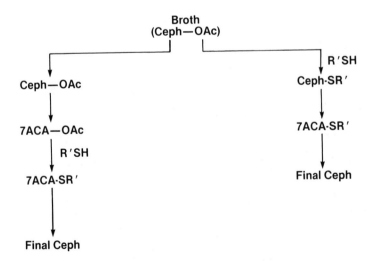

Cefamandole "Precursor"

Figure 11.   Example of displacement at C-3 prior to isolation of cephalosporin C.

Figure 12.   Displacement at C-3 prior to isolation of cephalosporin C.

Figure 14 depicts the final step in cephalosporin C synthesis as
being the enzymatic acetylation of desacetyl cephalosporin C with
acetylcoenzyme A. This step, however, only occurs intracellularly,
and exogenous desacetyl cephalosporin C is not reabsorbed (62,63).
       Chemical acetylation of the 3-hydroxymethyl to reform cephalos-
porin C is complicated by the competing lactonization reaction. This
problem is avoided by reacting a desacetyl cephalosporin C derivative
salt with acetic anhydride in a non-aqueous system. In the example
shown in Figure 15, the triethylamine salt of N-phthaloyl desacetyl
cephalosporin C is converted to N-phthaloyl cephalosporin C in
dimethylformamide. If the acetoxy group is to be subsequently displaced, other acid anhydrides can be used to prepare other C-3
esters (64). A variation of this approach, also shown in Figure 15,
is to react the desacetyl derivative with diketene, which yields
the acetoacetate ester (65).
       The extractive esterification method mentioned earlier allows
the desacetyl cephalosporin C derivative to be reacylated without
lactone formation since the C-4 carboxyl ester does not readily participate in the lactonization reaction. For the same reason, it is
possible to form a 3-halomethyl derivative of cephalosporin C from
desacetyl cephalosporin ester, with the resulting halide readily
displaced by the desired C-3 substituent. These reactions are shown
in Figure 16. (41)

Making a Choice

There is, of course, no one "right" way to process cephalosporin C
broth. It is a tribute to the purification chemists and engineers
that so many innovative approaches have been developed for cephalos-
porin C isolation. A decision to choose a particular route generally
involves weighing the following areas:
   • Capital costs
   • Processing costs
   • Throughput requirements
   • Yield potential
   • Product quality
   • Technical expertise available
   • Conformance to regulatory requirements
   • Waste treatment needs
   • Continuous or batch processing
   • Automation
   • Personnel safety and health
       What on the surface appears to be an attractive method may, on
closer analysis, turn out to have some serious flaws. As seemingly
straight forward as many of these approaches appear, when reduced
to practice it is not unlikely--in fact even probable--that some
complicating factors will be encountered. A certain amount of
commitment is needed, and even then one is reminded that what one
wants to do does not always match what is practical to do.

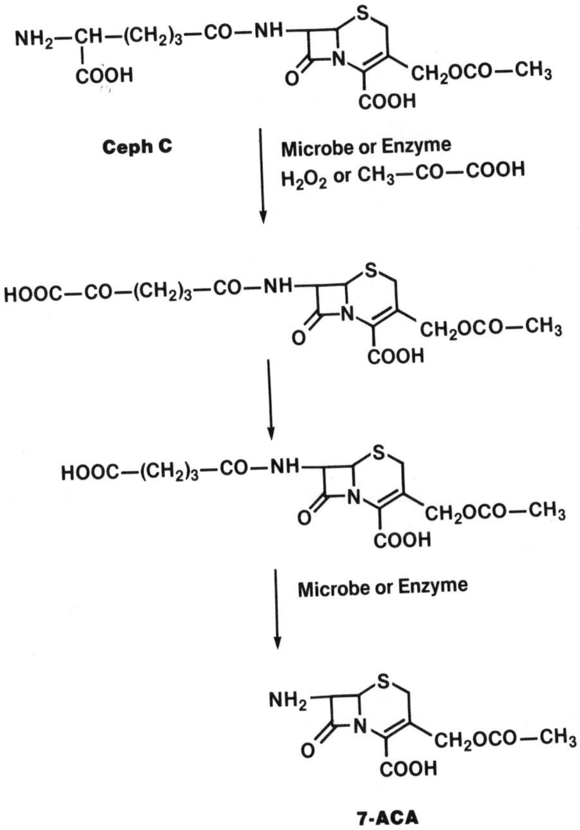

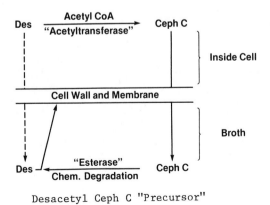

Figure 13.   Enzymatic or chemical deamination of cephalosporin C
and enzymatic conversion to 7ACA.

Figure 14.   Last step in biosynthesis of cephalosporin C and
exogenous derivation of desacetyl cephalosporin C.

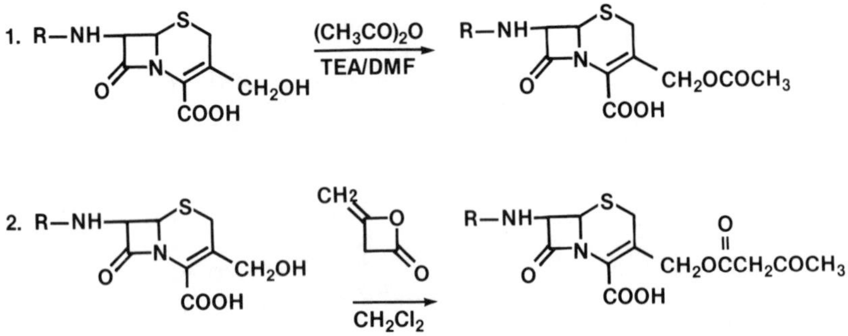

Figure 15. Non-aqueous acylation of 3-hydroxymethyl group of N-substituted   desacetyl cephalosporin C.

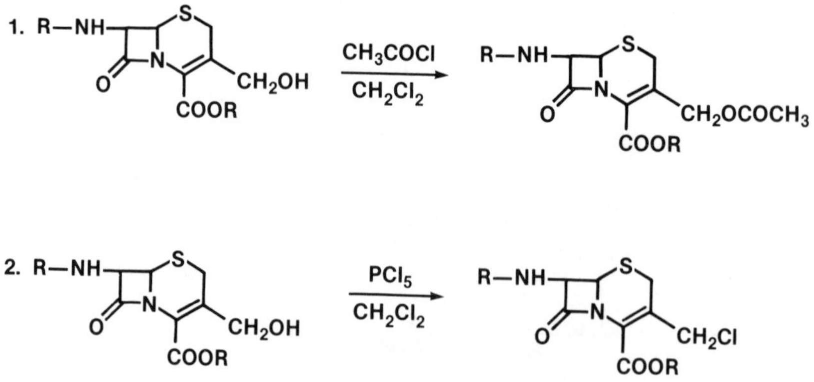

Figure 16. Examples of 3-hydroxymethyl reactions of 4-carboxyl esterified cephalosporin derivatives.

Literature Cited

1. Jeffrey, J.D'A.; Abraham, E.P.; Newton, G.G.F. Biochem. J. 1961, 81, 591-596.
2. Konecny, J.; Felber, E.; Gruner, J. J. of Antibiotics 1973, 26, 135-141.
3. Abraham, E.P.; Newton, G.G.F. U.S. Patent 3 093 638, 1963.
4. Takeda, Netherlands Patent 7 215 283, 1973.
5. Ciba, U.S. Patent 3 725 400, 1973.
6. Yamanouchi, Japanese Patent 79 11 298, 1979.
7. Meiji Seika, Japanese Patent 76 32 791, 1976.
8. Fujisawa, Japanese Patent 77 128 294, 1977.
9. Voser, W.; Weiss, K. J. Chromatogr. 1980, 201, 287-292.
10. Pirotta, M. Angew. Makromol. Chem. 1982, 109, 197-214.
11. Abraham, E.P.; Newton, G.G.F.; Hale, C.W. U.S. Patent 3 184 454, 1965.
12. Glaxo, German Patent 2 852 596, 1979.
13. Lilly, U.S. Patent 3 467 654, 1969.
14. Meiji Seika, Japanese Patent 77 76 486, 1977.
15. Merck, U.S. Patent 3 709 880, 1973.
16. Pfizer, German Patent 2 157 693, 1972.
17. Proter, Japanese Patent 78 112 892, 1978.
18. Lilly, U.S. Patent 3 853 863, 1974.
19. Meiji Seika, Japanese Patent 75 149 694, 1975.
20. Ciba, U.S. Patent 3 522 248, 1970.
21. Takeda, Japanese Patent 76 108 085, 1976.
22. Meiji Seika, German Patent 2 721 731, 1977.
23. Meiji Seika, Japanese Patent 78 53 689, 1978.
24. Meiji Seika, Japanese Patent 76 029 493, 1976.
25. Alfa, German Patent 2 507 117, 1975.
26. Takeda, German Patent 2 208 631, 1972.
27. Fujisawa, Japanese Patent 73 133 89, 1973.
28. Toyo Jozo, Japanese Patent 77 82 791, 1977.
29. LePetit, German Patent 2 458 554, 1975.
30. Bristol, U.S. Patent 3 573 296, 1971.
31. Meiji Seika, German Patent 2 418 088, 1974.
32. Lilly, U.S. Patent 3 641 018, 1972.
33. Alfa, Belgium Patent 796 540, 1973.
34. Sankyo KK, German Patent 2 523 280, 1975.
35. Lilly, U.S. Patent 3 980 644, 1976.
36. Glaxo, U.S. Patent 3 821 208, 1974.
37. Proter, British Patent 1 565 053, 1980.
38. Biochemie, British Patent 2 040 942, 1980.
39. Roussel-UCLAF, German Patent 2 841 363, 1979.
40. Andrisano, R.; Guerra, G.; Mascellani, G. J. Appl. Chem. Biotechnol. 1976, 26, 459-468.
41. Bywood, R.; Robinson, C.; Stables, H.C.; Walker, D.; Wilson, E.M. Spec. Publ. - Chem. Soc. 1977, 28, 139-144.
42. Glaxo, German Patent 2 436 772, 1975.
43. Ciba, U.S. Patent 3 661 901, 1972.
44. Lilly, U.S. Patent 3 835 129, 1974.
45. Bristol, German Patent 2 501 219, 1975.
46. Hoechst, German Patent 2 748 659, 1979.
47. Lilly, German Patent 2 054 085, 1971.
48. Bristol, Belgium Patent 836 428, 1976.

49. Fujisawa, Japanese Patent 71 14 735, 1971.
50. Asahi, Japanese Patent 81 53 688, 1981.
51. Asahi, U.S. Patent 4 079 180, 1978.
52. Yamanouchi, Japanese Patent 79 154 592, 1979.
53. Banyu, Japanese Patent 77 72 886, 1977.
54. Nippon KK, Japanese Patent 80 23 966, 1980.
55. Asahi, Japanese Patent 81 85 298, 1981.
56. Banyu, Japanese Patent 77 128 293, 1977.
57. Aries, French Patent 2 241 557, 1975.
58. Shibuya, Y.; Matsumoto, K.; Fujii, T. Agri. Biol. Chem. 1981, 45, 1561-1567.
59. Ichikawa, S.; Murai, Y.; Yamamoto, S.; Shibuya, Y.; Fujii, T. Komatsu, K.; Kodaira, R. Agri. Biol. Chem. 1981, 45,2225-2229.
60. Ichikawa, S.; Shibuya, Y.; Matsumoto, K.; Fujii, T.; Komatsu, K.; Kodaira, R. Agri. Biol. Chem. 1981, 45, 2231-2236.
61. Huber, F.M.; Baltz, R.H.; Caltrider, P.G. Appl. Microbiology 1968, 16, 1011-1014.
62. Fujisawa, Y.; Kanzaki, T. Agri. Biol. Chem. 1975, 39, 2043-2048.
63. Felix, H.R.: Neusch, J.; Wehrli, W. FEMS Microbiol. Lett. 1980, 8, 55-58.
64. Takeda, Japanese Patent 77 027 792, 1977.
65. Takeda, Japanese Patent 77 093 785, 1977.

RECEIVED September 7, 1984

Extraction of Insulin–Related Material and Other Peptide Hormones from *Tetrahymena pyriformis*

JOSEPH SHILOACH [1], CHAIM RUBINOVITZ [1], and DEREK LeROITH [2]

[1] Biotechnology Unit, Laboratory of Cellular and Developmental Biology, National Institute of Arthritis, Diabetes, Digestive and Kidney Diseases, National Institutes of Health, Bethesda, MD 20205
[2] Diabetes Branch, National Institute of Arthritis, Diabetes, Digestive and Kidney Diseases, National Institutes of Health, Bethesda, MD 20205

Traditionally, peptide hormones were thought to be synthesized and released only by vertebrate glandular tissues. More recently however, studies have demonstrated that nerves, cancers and other vertebrate tissues can produce hormonal peptides (1-3). In addition invertebrate tissues have been shown to contain many of these peptides. We have recently described the presence of hormone-like peptides in unicellular eukaryotes as well as in bacteria (4-8).

The amount of these recoverable materials from the microorganisms is very small, when compared to the production of enzymes or secondary metabolites such as antibiotics. Thus, to convincingly demonstrate the presence of hormonal peptides in microorganisms the fermentation of large quantities of cells in defined medium was required as well as improved large scale extraction and purification procedures. In addition, we utilized very sensitive and specific radioimmuno-assays and bioassays for detecting the low concentration of these materials.

This paper describes the various procedures developed and adapted for the extraction and identification of these hormonal peptides in Tetrahymena pyriformis.

METHODS

Growth and Harvesting of Organisms

Tetrahymena pyriformis (ATCC #30039) was grown in large volume 100 to 300 liters) fermentors in defined medium (9). The ambient temperature was kept constant at 27°C and agitation was maintained at 50 rpm with an aertion rate of 0.3 v/v/min. Growth was monitored by measuring the optical density of timed samples at 650 nm. The glucose concentration of these samples was determined by the glucose oxidase method (10). Dissolved oxygen concentration

was measured with a polarographic electrode (Instrumentation Lab)
installed in the fermentor. The fermentation was stopped at the
end of the logarithmic growth phase.

The cells were concentrated by passing through a Millipore
cassette using tangential filtration flow system which utilized 15
square feet of 0.45 μ membranes. The circulation rate was
maintained at 6 liters per minute with an input pressure of 8 psi
and an output pressure of 6 psi.

Insulin-Related Material

The classic acid-ethanol extraction procedure which is used for
extracting insulin from the pancreas of vertebrates (11), was used
to extract the insulin-related material from Tetrahymena pyriformis.
The concentrated cell suspension was homogenized in 5-10 volumes
of ice cold acid ethanol (0.2 N HCl/75% ethanol), using a Waring
blender. Following overnight mixing of the suspension at 4°C, the
supernatant was separated from the precipitate by centrifugation.
The ethanol constituent in the supernatant was evaporated. The
solution was then diluted with 5 volumes of acetic acid (1N) and
pumped onto Sep-Pak C_{18} (octadecasilyl-silica) cartridges (Waters
Assoc.). Material extracted from 10 g of cells was applied to
each 1 ml Sep-Pak cartridge. The C_{18} absorbed material was eluted
with a discontinuous ethanol gradient in 0.01 N hydrochloric acid.
Those fractions containing the insulin-like material were filtered
on Sephadex G-25 disposable columns (Pharmacia PD-10) to remove
the ethanol; the columns were eluted using 0.1N acetic acid. The
eluates were then lyophilized, pooled and gel filtered on a
Sephadex G-50 Column equilibrated with 0.05M $(NH_4)_2CO_3$. Each
fraction was lyophilized, reconstituted with water and aliquots
assayed for immunoreactive insulin (12). Fractions containing
insulin-like immunoactivity were further purified by a combination
of ion exchange chromatography and high performance liquid chromatography (HPLC). Those fractions with insulin-like immunoactivity were pooled and tested for biological activity (13).

Somatostatin-like Material

Acid ethanol, 10 vol, (0.2 N HCl/75% ethanol) was used to extract
the somatostatin-like material from the homogenized cells. The
suspension was mixed overnight at 4 °C followed by centrifugation.
The ethanol was evaporated and the remaining supernatant was
neutralized with NH4OH and boiled. The mixture was centrifuged
and the supernatant was applied to a Sep-Pak C_{18} disposable
cartridge (8) equilibrated in 0.1% TFA. The acetonitrile/TFA was
removed by lyophilization and each fraction was assayed for
somatostatin-like immunoactivity. The fractions with somatostatin
activity were pooled, and purified on HPLC and tested in a
bioassay (14).

ACTH-LIKE MATERIAL

The concentrated cell suspension was extracted in 10 volumes of
0.1M HCl/0.22M formic acid. The homogenate was defatted using

ethyl acetate/ether concentrated by lyophilization and centrifuged. The supernatant was then applied to Sep-Pak C_{18} columns
and the ACTH-related material absorbed to the C_{18} material was
eluted with 60% acetonitrile/0.01M trifluoroacetic acid. The
eluates were evaporated under reduced pressure, reconstituted in
0.1M HCl/0.22M formic acid and gel filtered on a Sephadex G-50
column. Each eluate fraction was assayed for ACTH-like
immunoactivity (15) and bioactivity (16).

Relaxin-Like Material

Five volumes of 1M HCl were used to extract the relaxin-like
material from the homogenized cells. The suspension was lyophilized and acetone added to a final concentration of 70%. The
pellet formed following centrifugation was discarded and the
acetone concentration in the supernatant was increased to 95% and
centrifuged again. The pellet that formed was then reconstituted
in water, dialyzed against 0.05M ammonium acetate buffer (pH 5.5)
and purified on a CM 52 cellulose column (7). The relaxin-like
material was eluted from the column with 0.05M ammonium acetate
containing 0.2 M NaCl. This was followed by further purification
on a Sephadex G-50 column in 6M guanidinum chloride followed by
HPLC.

Radioimmunoassays

Insulin-, somatostatin-, ACTH-, and relaxin-related materials were
detected in the partially purified extracts using specific double
antibody radioimmunoassays (12,8,16,7).

Bioassays

In addition to specific radioimmunoassays, bioassays were used to
measure the hormone-like molecules in the partially purified
extracts (13,14,16). Specificity of the bioassay was tested by
examining the biological activity of the materials in the absence
and presence of a specific antibody (4,5,8).

RESULTS

Tetrahymena Growth

The doubling time of the Tetrahymena during exponential growth was
in the range of 3-4 hours (Figure 1), when incubated at 27°C and
aerated at a rate of 0.3 v/v/min. At the end of the logarithmic
growth phase the glucose concentration decreased to below 0.2 g
per liter, and the percent saturation of dissolved oxygen ap-
proached zero.

The tangential filtration flow system concentrated the cell
suspension greater than 20 fold using 15 square feet of the 0.45 μ
membrane. The filtrate flow rate was 1 liter per minute and the
circulation rate was 6 l/min. Initial concentration of the cell

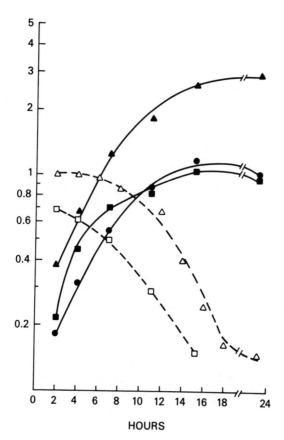

Figure 1. Large-scale fermentation of <u>Tetrahymena</u> <u>pyriformis</u>. The fermentor was inoculated with 10 liters of 48 hours <u>Tetra-hymena</u> culture. During growth, both oxygen and glucose concentrations fell. Key: ▲—▲, cells/ml 2 x 10^4; ■—■, cells gr/100 ml; ●—●, cells OD_{650} mμ; □--□, glucose gr/100 ml; △--△, dissolved oxygen % saturation.

suspension was 2.9 x 10^5 cells per ml and the final concentration was 6.4 x 10^6 cells per ml (Figure 2).

Insulin-related Material

Following the extraction of the cells in acid-ethanol, the material was applied to Sep-Pak C$_{18}$ cartridges and eluted by a discontinuous acid ethanol gradient. The insulin-related material was eluted in the 60% and 70% ethanol eluates (Figure 3).

Those fractions containing insulin-like material were pooled and the ethanol was removed by gel filtration on Sephadex G-25 disposable columns (Pharmacia PD-10) (17). The eluates were pooled and lyophilized. The material was then gel filtered on a Sephadex G-50 (fine) column and a peak of insulin-like immunoactivity was recovered in the region where mammalian insulins migrate on Sephadex G-50 (Figure 4).

Those fractions containing the immunoactive insulin were pooled, lyophilized and tested for insulin bioactivity in a bioassay which measures the incorporation of ^{3}H-glucose into toluene soluble lipids using adipocytes from male rats. (13). The amount of insulin bioactivity recovered from the extract was similar to the amount of immunoactivity obtained in the insulin radioimmunoassay. In addition the bioactivity was almost completely removed in the presence of an excess of anti-insulin antibody as well as by an antibody which blocks the insulin receptor (data not shown) (Figure 5) (4).

Insulin Specificity Studies

Since the concentration of the hormonal peptide was relatively small compared to the protein concentration in the extract, it was important to exclude the possibility that the results obtained in the radioimmunoassay and bioassay were due to non-specific interfering substances. Therefore, a number of control experiments were performed (Table I). These experiments had demonstrated that following 72 hours the ^{125}I-insulin tracer was not destroyed by incubation with the partially purified extract under the assay conditions described; further, the partially purified extract did not interfere with the interactions between tracer and the first (anti-insulin) antibody or with the interaction of first and second antibody. The neutralization of the insuln-like biological activity in the presence of the anti-insulin antibody as well as the anti-insulin-receptor antibody strongly suggested that the biological activity was acting via the insulin receptor on the cell surface of the rat adipocytes. Furthermore, it suggested that the insulin-like bioactivity was produced by material very closely resembling typical mammalian insulins (4).

Somatostatin-like Material

Following the partial purification of the acid-ethanol extract on Sep-Pak C$_{18}$ cartridges, the eluate fractions containing somatostatin immunoactivity were pooled, lyophilized and further purified on HPLC (Figure 6). The somatostatin-like material from

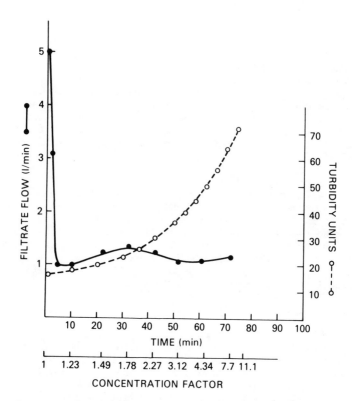

Figure 2. <u>Tetrahymena</u> concentration by tangential flow filtration. Batch size: 100 l. Area: 15 ft^2 (1.39 m^2). Circulation
rate: 6.0 l/min. P_{in}/P_{out}: 8/6. Final volume: 4.5 l. Concentration factor: 22. Flux: 0.79 l/m^2 min. Recovery: initial
cells no. 2.9 x 10^{10}; final cells no. 2.88 x 10^{10}.

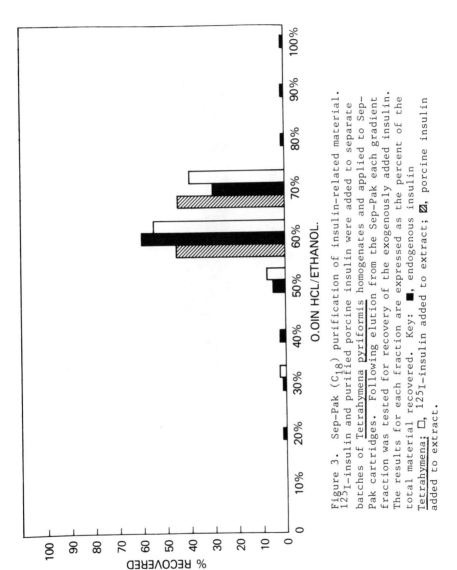

Figure 3.   Sep-Pak (C_{18}) purification of insulin-related material.
^{125}I-insulin and purified porcine insulin were added to separate
batches of <u>Tetrahymena pyriformis</u> homogenates and applied to Sep-
Pak cartridges.   Following elution from the Sep-Pak each gradient
fraction was tested for recovery of the exogenously added insulin.
The results for each fraction are expressed as the percent of the
total material recovered. Key: ■, endogenous insulin
<u>Tetrahymena</u>; ☐, ^{125}I-insulin added to extract; ▨, porcine insulin
added to extract.

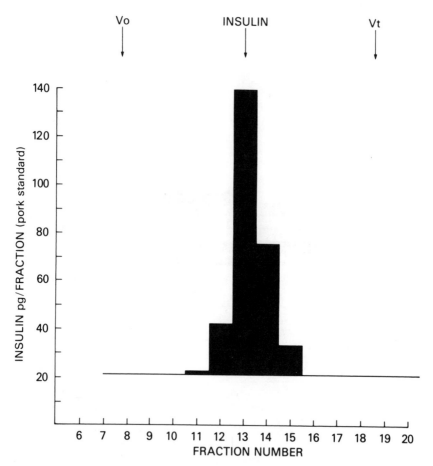

Figure 4. Sephadex G-50 filtration of Tetrahymena pyriformis
extract. Results are plotted as insulin immunoactivity in each
fraction using a porcine insulin radioimmunoassay. The horizontal
line represents those fractions tested and the lower level of
sensitivity of the assay. The left arrow (Vo) represents the
void volume, the right arrow (Vt) represents the salt elution
position, and the middle arrow represents the region where ^{125}I-
insulin and purified porcine insulin elutes on the same column.
(Reproduced from Ref. 4.)

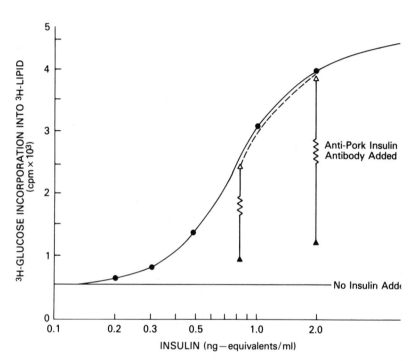

Figure 5. Insulin-related bioactivity of the extract of Tetrahymena pyriformis. Key: ●, pork insulin standard; △, Tetrahymena extract; ▲, Tetrahymena with anti-insulin antibody added.
(Reproduced from Ref. 4.)

Table I. Exclusion of Nonspecificity in the Assays

RADIOIMMUNOASSAY

Test	Result	Conclusion
1) Precipitation of ^{125}I-insulin with 10% TCA after incubation with partially purified extract.	>90% of ^{125}I-insulin precipitated	Extract did not degrade the tracer
2) Sephadex G-50 gel filtration of ^{125}I-insulin before and after 72 hours of incubation with extract	Elution profiles were identical	Extract did not degrade tracer; Absence of possible tracer binding substance
3) Excess anti-insulin antibody (1st antibody) added to extract; 2nd antibody then used to precipitate the antigen-antibody complex	>90% of the ^{125}I-insulin precipitated by the antibody	Extract did not interfere with the interaction of tracer with anti-insulin antibody; nor with the first and second antibody interactions
4) ^{125}I-insulin incubated for 3 days with extract; then the ^{125}I-insulin is diluted and used as tracer in a radioimmunoassay	Normal standard radioimmunoassay obtained	No change in ^{125}I-insulin affinity as a result of contact with the extract
5) Tetrahymena extract was added to RIA for hGH using the same 2nd Ab.	No interference in the hGH assay	Extract has no effect on the 2nd Ab

BIOASSAY

Test	Result	Conclusion
1) Addition of anti-insulin receptor antibody	>90% of the bioactivity was removed	Suggests that the purified extract is active through the insulin receptor on the surface of the rat adipocyte
2) Addition of anti-insulin antibody	>80% of the bioactivity was neutralized	Suggests that the bioactivity is produced by materials very similar to the typical mammalian insulins

SOMATOSTATIN

Hydrochloric acid-ethanol
|
Sep-Pak C$_{18}$ reverse phase
|
HPLC
/ \
RIA   Bioassay

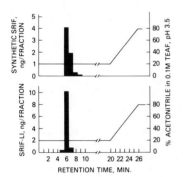
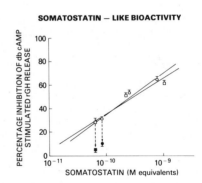

Figure 6. Somatostatin-like material in Tetrahymena pyriformis. Key (right panel): □, synthetic somatostatin; O, Tetrahymena extract (HPLC purified); ■ and ●, anti-somatostatin antibody added. (Reproduced from Ref. 8.)

the partially purified Tetrahymena extract eluted with a retention time very similar to that of synthetic somatostatin run under the identical conditions. Following HPLC, the somatostatin-like material was tested for biological activity in a bioasssay which measures the suppression of growth hormone (GH) release from isolated rat pituitary cells. Tetrahymena somatostatin-like material inhibited GH release and the bioactivity was equivalent to the amount of immunoactivity. In addition the bioactivity was neutralized in the presence of anti-somatostatin antiserum (Figure 6).

ACTH-like Material

Sephadex G-50 gel filtration revealed multiple partially resolved peaks of immunoactive ACTH (Figure 7). The most prominent peak of ACTH immunoacativity eluted in the region where hACTH (1-39) standard elutes under similar conditions. The fractions containing this peak of ACTH immunoactivity were pooled, purified on SDS polyacrylamide gel electrophoresis (data not shown) and further tested in a bioassay measuring corticosterone release from isolated adrenal cells from rats (Figure 7). The biological activity produced by the partially purified Tetrahymena extract demonstrated an increase parallel to that given by the hACTH standard. Furthermore, the biological activity was neutralized by anti-ACTH antiserum (Figure 7).

Relaxin-Related Material

Following the purification of the Tetrahymena extract on CM-cellulose chromatography and Sephadex G-50 gel filtration (7), the relaxin-like material was further purified on HPLC (Figure 8). The relaxin immunoactivity eluted 2-3 minutes earlier than purified porcine relaxin. The HPLC purified relaxin-like material was tested for the presence of disulfide linkages in the molecule by reduction and alkylation. The purified extract was reduced under nitrogen with an excess of dithiothreitol for 1 hour, followed by the addition of a 10 fold excess of iodoacetic acid. A parallel experiment was performed using purified porcine relaxin and samples were tested for immunoactivity before and after reduction and alkylation. Reduction and alkylation caused a marked decrease in both Tetrahymena and porcine relaxin immunoactivities (Figure 8). These results strongly suggest that Tetrahymena relaxin-like material contains disulfide bridges.

Table II. Materials Resembling Vertebrate Hormonal Peptides
            Extracted and Purified from T. Pyriformis

Hormonal Peptide	Immunoactivity
	(pg equivalents per g wet weight cells)
Insulin	20-40
Somatostatin	10-20
ACTH	500
Relaxin	2,200

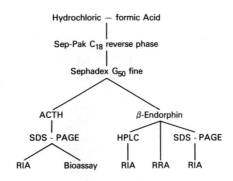

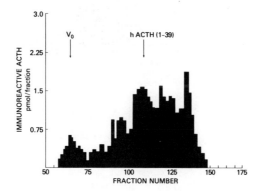

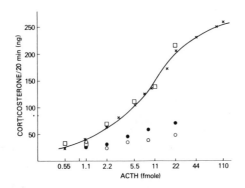

Figure 7. ACTH-like material in Tetrahymena pyriformis. Key (bottom panel): □, the extract; x, synthetic human ACTH (1-39) standard; O, anti-ACTH antiserum; ●, Sepharose-protein A immunodepletion. (Reproduced from Ref. 5).

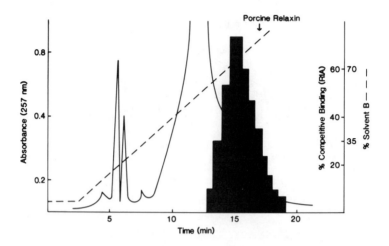

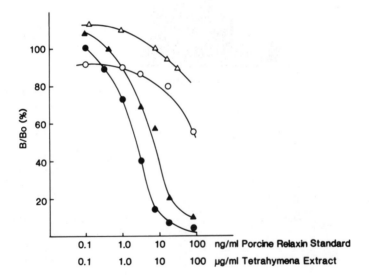

Figure 8. Physico-chemical characteristics of relaxin-like material in <u>Tetrahymena</u> <u>pyriformis</u>. Key: ● and ○, <u>Tetrahymena</u> relaxin; ▲ and △, porcine relaxin. (Reproduced from Ref. 7.)

DISCUSSION

We have previously demonstrated the presence of peptides resembling insulin, somatostatin, ACTH, and relaxin in Tetrahymena,
unicellular fungi and prokaryotes (18,19). This paper describes
some of these findings and emphasizes the important biotechnical
techniques used in these studies.

Since the concentration of the hormonal peptides in unicellular oganisms is low (pg per gr cells) it was necessary to
grow up large amounts of cells. Thus, 100-300 liter fermentors
were utilized for growing Tetrahymena pyriformis in defined
medium. This required careful control of fermentation conditions
paying particular attention to aeration, oxygenation, pH and
glucose concentration. After the logarithmic growth phase the
growth of Tetrahymena slowed with a decrease of the glucose
concentration in the media; this represents a growth profile
typical of microorganisms. At this point the dissolved oxygen
concentration was approaching zero indicating that at this rate of
oxygen supply the cells had depleted most of the dissolved oxygen.
Due to the fragility of the cells at this phase of the growth
cycle, we did not increase the air supply or the agitation of the
cells and medium. It was important to keep the cells intact
since their disruption would release the small quantities of
hormonal-like material into a large volume making concentration and
extraction a formidable task.

Harvesting of the cells at the end of the logarithmic growth
phase was found to be particularly difficult. Due to the fragility of the cells, continuous centrifugation systems proved to be
inadequate. Therefore, tangential filtration flow was used.
Despite the high flow rate through the cassette system, a 20 fold
concentration of cells was attained with minimum disruption.
Furthermore, the Millipore cassette system was capable of concentrating the large volumes at a constant rate of 50 liters per
hour. This rapid concentration rate also enabled the harvesting to
be completed quickly thus minimizing the enzymatic degradation of
the hormone-like peptides.

Small polypeptides and proteins are generally soluble in acid
solutions. Thus cold acids have been used to extract insulin,
somatoststin, ACTH and relaxin from vertebrate tissues. We used
similar solutions to extract these peptides from Tetrahymena
pyriformis cells. Unfortunately, losses of the peptides in
utilizing these extraction procedures as well as the purification
methods occurred. By using tracer labelled peptides and purified
mammalian peptides we have previously estimated our recoveries
during the extractions and purification to be about 10-30% (17).
The purification methods utilized in these studies were adapted
from previously described techniques for purifying vertebrate
hormones from glandular tissues. We are at present studying other
methods to improve our recoveries.

Despite the extremely low concentration of the hormone-like
peptides in the microorganisms, we were able to detect them by
very sensitive radioimmunoassays and bioassays which had previ-

ously been developed for very low circulating levels of these hormones. Thus we could detect pg levels of these hormones. One problem which was rigorously tested was the question of non-specific interaction in the radioimmunoassays and bioassays (see Results, Table I). Having excluded the possibility of non-specificity we addressed the question of whether the hormone-like activities found in Tetrahymena were due to exogenous contamination of the extract or were indeed native to the organisms.

Numerous experiments were performed to exclude possible contamination of the extracts from exogenous sources, these included;

1. Unconditioned medium was processed in the fermentor, extracted and purified in an identical manner to that used for the cells and was found to be devoid of any hormonal immunoactivity.
2. Tetrahymena cells were grown in a separate laboratory under identical conditions and were shown to contain similar amounts of insulin immunoactivity (17).
3. Time course experiments demonstrated a time dependent increase in insulin-like material extracted from both cells and medium (17).
4. Subcellular distribution of endogenous T. pyriformis insulin-like material differed from the subcellular distribution of ^{125}I-insulin and purified porcine insulin which had been added to Tetrahymena cells prior to subcellular fractionation (17).
5. When endogenous insulin-like material was destroyed by acid-hydrolysis after homogenization of the cells no insulin was detected following the purification of the extracted cells (17).

The presence of typical vertebrate hormonal peptides in unicellular eukaryotes and prokaryotes, suggest that the endocrine and nervous systems of multicellular organisms have evolutionary origins much earlier than previously thought.

Finally, we have demonstrated the presence of these hormonal peptides in microorganisms by their activity in various assays. In the future we would like to get more direct evidence by obtaining enough material for direct analysis and to try and identify the DNA region responsible for the insulin expression.

Literature Cited

1. Krieger, D.T.; Martin, J.B. New Eng. J. Med. 1981, 304, 944-51.
2. Hokfeldt, Tl; Johansson, O.; Ljungdahl, A.; Lundberg, J.M.; Schultzberg, M. Nature 1980, 284, 515-21.
3. Odell, W.D.; Wolfsen, A.R. Am. J. Med. 1980, 68, 317-18.
4. LeRoith, D.; Shiloach, J.; Roth, J.; Lesniak, M. Proc. Natl. Acad. Sci. USA 1980, 77, 6184-88.
5. LeRoith, D.; Liotta, A.S.; Roth, J.; Shiloach, J.; Lewis,M.E.; Pert, C.B.; Krieger, D.T. Proc. Natl. Acad. Sci. USA 1982, 79, 2086-90.
6. Le Roith, D.; Shiloach, J.; Roth, J.; Lesniak, M. J. Biol. Chem. 1981, 256, 6533-36.

7. Schwabe, C.; LeRoith, D.; Thompson, R.P.; Shiloach, J.; Roth, J. J. Biol. Chem. 1983, 258, 2778-81.
8. Berelowitz, M; LeRoith, D.; VonSchenk, H.; Newgard, C.; Szabo, M; Frohman, L.A.; Shiloach, J.; Roth, J. Endocrinology, 1982, 110, 1939-44.
9. Holz, G.G.; Erwin, J.; Rosenbaum, N.; Aaronson, S. Arch. Biochem. Biophys. 1962, 98, 313-22.
10. Raabo, E.; Terkildsen, T.C. Scand. J. Clin. Lab. Invest. 1960, 12, 402.
11. Mirsky, I.A. In "Methods in Investigative and Diagnostic Endocrinology"; Berson, S.A., Ed.; American Elsevier: New York, 1973; pp. 823-993.
12. Yalow, R.S.; Berson, S.A. J. Clin. Invest. 1960, 399, 1157-75.
13. Moody, A.J.; Stan, M.A.,; Stan, M.; Gliemann, J. Horm. Metab. Res. 1974, 6, 12-16.
14. Szabo, M.; Berelowitz, M.; Pettengill, O.S.; Sorrenson, G.D.; Frohman, L.A. Clin. Endocrinol. Metab. 1980, 51, 978.
15. Liotta, A.S.; Krieger, D.T. J. Clin. Endocrinol. Metab. 1980, 40, 218-77.
16. Liotta, A.S.; Krieger, D.T. Endocr. Res. Commun. 1977, 4, 158-9.
17. LeRoith, D.; Shiloach, J.; Heffron, R.; Rubinovitz, C.; Tenenbaum, R.; Roth, J. Submitted for publication.
18. Roth, J.; LeRoith, D.; Shiloach, J.; Rubinovitz, C. Clin. Res. 1983, 31, 354-63.
19. Roth, J.; LeRoith, D.; Shiloach, J.; Rosenzweig, J.; Lesniak, M.A.; Havrankova, J. New Eng. J. Med. 1982, 306, 523-7.

RECEIVED October 12, 1984

Author Index

Subject Index

A

193

Production by Meg Marshall
Indexing by Karen McCeney
Jacket design by Pamela Lewis

Elements typeset by Hot Type Ltd., Washington, D.C.
Printed and bound by Maple Press Co., York, Pa.